T0331935

MATCHMAKERS AND MARKETS

MATCHMAKERS AND MARKETS

The Revolutionary Role of Information
in the Economy

Yi-Cheng Zhang

OXFORD
UNIVERSITY PRESS

OXFORD
UNIVERSITY PRESS

Great Clarendon Street, Oxford, OX2 6DP,
United Kingdom

Oxford University Press is a department of the University of Oxford.
It furthers the University's objective of excellence in research, scholarship,
and education by publishing worldwide. Oxford is a registered trade mark of
Oxford University Press in the UK and in certain other countries

First Edition published in 2020

Impression: 1

Published in the United States of America by Oxford University Press
198 Madison Avenue, New York, NY 10016, United States of America

British Library Cataloguing in Publication Data
Data available

Library of Congress Control Number: 2019947650

ISBN 978–0–19–884098–5

Printed and bound by
CPI Group (UK) Ltd, Croydon, CR0 4YY

Foreword by Berno Buechel

This book starts by criticizing the kind of economics that I regularly teach to undergraduates, and in some cases to graduate students. Instead of simply offending traditional economics, it provides a healthy criticism that opens the eyes to blind spots and, more importantly, provides an alternative view on how markets work. This view is novel and exciting, and convincingly explains several phenomena that emerge in the digital economy. I highly enjoyed reading this book. I believe that it will have a deep impact on me, as well as on many of you.

When I first met Yi-Cheng Zhang, I aimed to bring together his knowledge as a renowned physicist and my knowledge as an economics professor, to collaborate on the emerging field of *network science*. Little did I know that he would soon be teaching me economics.

Digitalization is transforming the economy in innumerable ways. This leads to phenomena that traditional models fall short of explaining. This book manages to elegantly explain several of these. Consider the rise of several Internet giants whose businesses can be described as matchmaking or operating two-sided platforms. Only recently have economists started to analyse such two-sided platforms. Their focus is often on strategic price setting, whereas Prof Zhang emphasizes the fundamental importance of information. And while economic models are often static and somehow restrictive, he widens the view for *an eco system of information that is dynamic and open*. Hence the approaches are complementary, and there is indeed much to learn from each other.

Understanding the value of rigorous theory and having open eyes for the complexity of life are not two opposing forces. On the contrary, it is their combination that helps us deepen the reflection on a topic, and it clearly does so for the theoretical physicist who wrote this book. It can also work for economists and everyone else. As an economist specializing in network science, I have seen a similar turning point. For a long time, economists had analysed decisions of consumers and firms without taking into account how they are embedded in a network of connections. When social networks became a flourishing field in economics during the last two decades, there were already rich strands of literature in various disciplines—from sociology to physics—that could be (re)discovered. Economics can tremendously benefit from such a fruitful exchange with other disciplines, and it offers its own insights in return.

This book contains remarkably innovative views, and is filled with wisdom. The content is based on solid scientific grounds, while the presentation makes it easy to access and fun to read. I also use it for teaching. For instance, in my graduate course 'Digital Economy' I want my students to understand the different roles of matchmakers and the importance of recommender systems, for which this book is also enlightening. To undergraduate students in the 'Microeconomics' course it illustrates, among other concepts, how restrictive it is to assume that a consumer knows all available products. I highly recommend this book to students of management and economics around the world, as well as to the practitioners in Silicon Valley and other regions. For everyone interested in the digital economy, I consider it a must-read.

Prof Zhang is a renowned scholar in physics and beyond. The frequency that his work is cited and his role for the company Alibaba are just simple indications of the highest respect that he receives in both the academic world and the business world, and I am proud to call him my colleague and friend. Being neither billionaire nor Nobel Prize laureate (not even close!), I feel honored to contribute this Foreword. I hope that you will all share my enthusiasm for this book and I wish you a wonderful journey of reflection.

Berno Buechel is a professor of economics at the University of Fribourg (Switzerland) and co-founder of the Swiss Center for Data + Network Sciences. His research focuses on the working of markets on strategic behavior and on the role of social networks.

Foreword by Jean-Philippe Bouchaud

In the wake of the 2008 Great Recession, critics of mainstream economics have become more and more vocal. Whereas Robert Lucas argued that "the crisis was not predicted because economics theory predicts that crises cannot be predicted," Willem Buiter, in one of the most insightful and damning pieces I have read, wrote that "most mainstream macroeconomic theoretical innovations since the 1970s have turned out to be self-referential, inward-looking distractions at best. Research tended to be motivated by the internal logic, intellectual sunk capital, and aesthetic puzzles of established research programmes rather than by a powerful desire to understand how the economy works—let alone how the economy works during times of stress and financial instability."[1]

The search for a new paradigm in fact predates the crisis, with many luminaries (H. Minsky, R. Shiller, R. Thaler, and D. Kahneman, among others) insisting on the importance of behavioral biases and of feedback loops that jeopardize the pillars on which economic theory has been resting in the past fifty years. Although these ideas have progressively made their way into the mainstream, their status is still ambiguous: you need a model to beat a model. Unfortunately, there is at this stage no consistent framework that can compete with the powerful methods of classical economics, which deals with a world where agents face a series of well-posed optimization problems for which all constraints are known and all states of the world are identified. Markets are perfect and reveal all known unknowns and all probabilities. The only problem is to compute the optimal solution. All other contingencies are taken care of by the magic of the invisible hand. As F. Hahn has said: "We need not worry about exhaustible resources because they will always have prices which ensure their proper use."[2]

In the present inspiring and deeply original essay, driven by a powerful desire to understand how the economy works, Prof Zhang explains how the world we live in fundamentally differs from the one modeled by economists. This is particularly true of our post-Internet world, where information plays an ever-growing role. The *quantity* of stuff that we produce—an obsession born during war times—becomes secondary to the *quality* of the stuff we consume, of our leisure time and of our quality of life. In fact, quantity becomes irrelevant when our very survival and that of our environment is at stake. But these aspects are nowhere encoded in the "utility functions" that firms and households are supposed to optimize. No wonder little progress can be made based on traditional cost–benefit analyses.

Prof Zhang argues that it is just not enough to recognize that humans are prone to behavioral biases, and that imperfections and frictions abound in a world otherwise well captured by the framework of classical economics. Within this paradigm, imperfections merely prevent the system from settling into the ideal equilibrium of economic theory.

But a more radical departure is needed, Prof Zhang writes, where this ideal equilibrium does not even exist. The very concept of "equilibrium" does away with the fact that economy is always in flux, that unpredictable innovations can totally disrupt the economic landscape. What people actually think and do impacts the economy and changes the very parameters of the optimization problem that agents are supposed to solve. The subprime crisis should have been a minor shock. Instead, it spiraled into a self-fulfilling trust collapse. Consumers postponed purchases and firms postponed investments, driving the economy into the Great Recession. Imperfections are not a small perturbation to an otherwise ideal world; they are, on the contrary, central to both creative and destructive disruptions. We need a synthesis between Schumpeter and Soros, perhaps with a tinge of Andrew Lo and Nassim Taleb, and inputs from physics, biology, and computer science.

Prof Zhang is one of the most creative statistical physicists of his generation. He is primarily known for the so-called KPZ equation that, together with Mehran Kardar and Giorgio Parisihe, he invented in 1986 to describe random surface growth.[3] This triggered an enormous amount of activity both in the physics and mathematics communities, crowned by Martin Hairer's Fields Medal in 2014. His seminal *Minority Games* counts as one of the real successes of the nascent field of econophysics.[4]

The present book was in the making for many years, and many of us were eager to read Prof Zhang's insights. The result came as a kind of surprise: far from being a technical book promoting conceptual models that may compete with those of classical economics, this book focuses on the big picture, sowing ideas that need to be developed into a proper theory. Zhang the physicist seems to have been superseded by Zhang the philosopher, strongly influenced by his double Chinese–Western culture and by exposure to the information economy through his tenure at Alibaba. His insights on recommendation systems, on the value of information, and on the new types of information asymmetry are particularly relevant and noteworthy.

In truly inspiring books there are many more questions than answers. I certainly hope that some of the paths opened in this book will lead to important developments in the future—maybe by Prof Zhang.

Preface

In the final decades of last century, physics, especially theoretical physics, attracted many young students, and I was among them. It was an exciting period, as amazing discoveries ranging from the subatomic world to the universe came in succession. About the turn of the century, physics seemed to be a relatively mature science, and a crowd of theoretical physicists looked to other interdisciplinary fields for new excitement. Obviously, physicists' new love for complexity is no accident: Steven Hawking has famously predicted that the twenty-first century would be the "century of complexity," instead of the century of biology or physics.

In theoretical physics we were accustomed to getting results fast, but little did I know that what I took on as a side journey wandering into finance and economics would turn out to be a much harder nut to crack. Just one problem has absorbed me for more than twenty years, and it is far from being completed. Had I known that economic problems are so hard, I am not sure that I would have had the courage to do it again!

If people find mainstream economics unsatisfactory, I believe that it is not that economists have done a bad job, but that theirs is a much harder science compared to physics. To understand, for example, financial markets with thinking agents is far more challenging than to study a gas of obedient atoms.

The challenge is still ongoing, but a first coherent picture has already emerged which allows me to present my investigation in a book for the public. During my ordeal, many disjointed analyses and isolated insights suddenly appeared in a coherent structure following the main theme on the role of information in the markets. I shall not dwell on my untold many hesitations and wrong turns that are buried in many yellowed notes. Over the years I have also released several earlier versions of this book in various formats for small circulation.

During the past two decades I have had the good fortune to interact with many economists, business leaders, startup entrepreneurs, and data scientists. About a decade ago, at the instigation of Mr Chunxiao Liang, then the senior vice president of Alibaba Inc, I helped establish an entity called Alibaba Research Center for Complexity Sciences to study information's role in consumer markets, and I have enjoyed privileged access to many Alibaba top executives, including Jack Ma. I could closely observe how Alibaba enabled millions of consumers and vendors to do ecommerce, and talk to many frontline people in various divisions. I have learned much from them, not only of the ecology of ecommerce, but also of their views on local and global competition, current and future.

My new theory is not a direct confrontation against mainstream economics; rather, it starts from a similar ground with the supply–demand relation amended with the explicit role of information, then covers what is omitted by mainstream economics. After all, economics was conceived in another age much ahead of the current economy. Since then, Information Technology has enabled many new business models which have transformed the markets in many ways that could not be foreseen more than a century ago. Now it is also time to transform our understanding of the economy based on a modern perspective.

In the mainstream theory of consumer markets, it is still about dual relationships—buyers and sellers—and the information's role, if any, is an afterthought. In a sharp departure from previous theories, this book deals with three-way relationships: buyers, sellers, and information intermediaries (matchmakers). Three-way relationships herald a whole new perspective regarding the markets, and the book's title represents this glaring contrast.

I must explain why the word "matchmaker" appears so prominently in this book. Traditionally, matchmaking is about how to make marriage-minded men and women pair with each other. Many people get married by finding their partners without the service of matchmakers, but a small fraction of people do use a matchmaker's service. Using marriage as a metaphor, I consider consumers and businesses as the potential mates to be paired. We may have the impression that they find each other by themselves, but if we look closer we may realize that the majority of the deals are facilitated by third parties. There is a broad category of such third parties that I generically call "information matchmakers", who in one way or another enable the two sides to find each other. In today's markets the information load is so overwhelming, both for consumers and businesses, that it is almost impossible for them to completely forego the service of such matchmakers. While for the marriage market the service of matchmakers is often unnecessary, we shall see that in consumer markets matchmakers are ubiquitous, as most of them are uninvited and hence remain invisible. Collectively, the visible and invisible matchmakers are vital for our consumer markets and beyond. In a broader sense, we shall see that the concept of matchmakers extends to other institutions facilitating all types of positive-sum games; indeed, the win–win–win proposition is the main theme throughout this book.

It may seem that merely adding information to the supply–demand relationship is an academic task, but for a systematic account of information's role in the markets we need to consider agents' information capabilities, and these depend crucially on how matchmakers operate. We shall see that once we add these third parties, other questions arise naturally, such as information ecology in the markets, human agents' motivations, connections, social structures, and institutions, all of which are part and parcel of a viable theory of markets.

I expect that readers will be able to relate their own experience and knowledge and use this book as a stepping stone to ponder small and large issues of our consumer markets and beyond.

I wish to acknowledge the following people for their criticism, encouragement, and help. Brian Arthur, (the late) Per Bak, Jean-Philippe Bouchaud, Berno Buechel, Damien Challet, Yongchao Duan, Steve Keen, Chunxiao Liang, Linyuan Lv, Matus Medo, Andrzej Nowak, Paul Omerod, Luciano Petronero, Brigitte Rosewell, Canzhong Yao, Wuyang You, Ming Zeng, Lei Zhou, and Tao Zhou. I must acknowledge the support of my family during the labor on this book: father Yingtai, mother Xueying, wife Tetyana, daughter Gaia Nicole, and brother Jun.

Contents

Introduction

The world economy has seen dramatic changes since the Industrial Revolution, and its focus has now shifted from smokestack to information. Mainstream economics has been the dominant theory, and the growing mismatch between theory and the real world calls for alternative explanations. Instead of offering one more critique, this book aims to build an alternative economic theory in which information plays the central role.

Information does not figure prominently in mainstream textbooks. The key parameters are prices and quantities, and people are assumed to know the quality of products. This is a far cry from the real world in which a consumer faces many choices, and it is far from easy to know the quality and suitability of products. On financial markets, information limitations are even more severe. A stock cannot simply be represented by its price, earning ratio, and a few other parameters, as there are numerous idiosyncratic details for each stock that investors must investigate.

At the core of mainstream economics is the paradigm of allocation of scarce resources, which aims at optimization under fixed constraints. Princeton economist Dani Rodrik, in his popular blog,[1] succinctly described mainstream economics: "Here is a bunch of firms, here are their choice variables, here is the market structure under which they operate, here is what they maximize, and here is what the equilibrium will look like." Indeed, if the variables and constraints were given, the only reasonable thing to do is maximization, and economic problems would indeed reduce to mathematical exercises such as those we often see in economics textbooks and literature.

While Rodrik's statement is plausible, we shall show that it misrepresents the real world. Facts cannot simply be labeled as variables, since they are known to only a limited number of people, and each of those who do know has a partial understanding. To achieve any goal (for example, maximization) the process takes time, during which the market constraints rarely remain fixed. Constraint-shifting invalidates the mainstream methodology, ceteris paribus. In sharp contrast, our theory presents a dynamic approach that regards the constraints as movable; instead of the final state we shall focus on endless processes and investigate their causes and consequences.

Our theory is based on a new paradigm that combines allocation and creation of resources. The economy is full of allocative and creative actions, and the role of competitors (allocative) and that of innovators (creative) often mingle. The allocation paradigm focuses on optimization under constraints, and the allocation-creation paradigm studies why and how the constraints shift. In the real world, most processes towards a perceived goal will redefine the constraints.

Resources creation is not confined in a black box. Lionel Robbins once told George Soros,[2] then his student, that the task of economics is to study the relation between supply and demand; economists should not probe what is underneath. The new paradigm will focus on the underneath, since explicit variables and implicit knowledge are intimately connected.

For material resources, the more that are used the fewer remain; for economic opportunities in the twenty-first century, the more that are exploited the still more may be revealed and created. Instead of diminishing returns, the new paradigm considers an open evolution leading to an ever more complex world filled with unexpected opportunities and concomitant risks. Economic history is not one of repeated optimization, and wealth does not emerge by magic. How we allocate current resources will impact on future resources creation; in the economy, we can never have a fixed box within which we solve optimization problems.

Mainstream economics portrays consumer markets as having two types of participants: buyers and sellers. But in advanced economies, buyers and sellers do not sufficiently know the other side, hence we speak of ubiquitous information deficiency and the necessary rise of information intermediaries (matchmakers). Most of intermediaries, however, are not recognized as such, as they often perform the matchmaking role on the side. When the two sides (consumers vs. businesses) are joined by third parties, the three-way relationship gives rise to many fascinating and complex scenarios, we can only outline the main features in this book. Third parties appear in myriad disguises and span the whole spectrum from consumers to businesses. Our methodology allows us to study their strategies and evaluate the effects.

This book's key defining feature is the view of markets as three-way relationships vs. the mainstream view of dual relationships. Here we present the layout of this book, and also see why the subject warrants a systematic discussion far beyond the markets themselves and why one chapter calls for the next as the main logic develops.

Strictly speaking, if we want a new account of consumer markets highlighting information's role to the supply–demand law, Chapter 1—a narrative summarizing a dozen published papers—would suffice. Why do we need the following nine chapters? They need an apologia to justify the book's length.

In Chapter 1 we show that information capabilities can impact market transactions. Instead of just waiting for these capabilities to rise or fall as a fatality, we show that there are means and motives to change them. But consumers cannot handle the information burden alone. Third parties can either help or exploit them. Chapters 2 and 3—one focusing on the present and the other on the near future—discuss the role of information intermediaries or matchmakers, and how third parties can play a pivotal role in consumer markets. Chapter 3 is actually a prediction of an enabler of future consumer markets.

Chapter 4 discusses the consequences of Chapters 1–3: that enhanced information capabilities lead to product diversification. If this assertion turns out to be valid, then economic growth will not only be measured by GDP increases but also by product diversification.

Chapter 5 and 6 extend the same logic to two other markets: finance and information content. Besides the two important markets per se, by discussing their common features with the consumer markets we may gain insights into the common theory and the detailed mechanisms of different types of markets.

But we cannot stop after these three types of market. In Chapter 7 we discuss how markets and the economy evolve. People are often tempted to compare economic evolution with Natural Evolution. Just as Natural Selection is the agent for biological evolution, here we identify informational selection as the agent for economic evolution. Whereas the basic principles are similar to Darwinian theory, we show that economic evolution has many important particularities. The detailed mechanisms may stimulate diligent readers to ponder deeper questions concerning the economy as a whole.

Chapters 8 and 9 discuss human motivations and the implications for society. These are inspired by psychology and social studies. We posit that the ultimate source of economic growth is from the ever-increasing wants and skills of human agents, that informational selection picks them up and institutions connect them, so that one fraction of these wants and skills convert into supply and demand.

Chapter 10 summarizes the main theme of this book by noting that we actually follow a new paradigm that may again compare with the allocation paradigm of mainstream economics. In a way, only Chapters 1 and 10 afford a direct comparison with mainstream theory. The former establishes an amended supply–demand relation, and the latter summarizes the new methodology. Chapters 2–9 concern the details of how the new paradigm might work in the real world.

By reading this book, entrepreneurs, financiers, policymakers, and the public at large might relate their own experiences and insights and reflect on their own role in the dynamic, vibrating economy. They may either refute the conclusions proposed in this book, or improve them, or add their own.

PART 1

INFORMATION AND CONSUMER MARKETS

1

Magic Pie

This first chapter lays out the basic tenants of our new market theory. Based on the recognition of widespread information asymmetry in consumer markets, consumers have, only a limited understanding of almost all products. The gray-degree of this extent plays a key role. Consumers and businesses have motives and means to shift the gray-scale, and this can impact market transactions in the new supply–demand relation. The key concept is the fundamental asymmetry between consumers and businesses, that businesses are easier to adapt to consumers' needs rather than vice versa. Hence we advocate finding ways to improve consumers' selection power incessantly as the way of value creation.

1.1 Information capabilities

The role of information in consumer markets cannot be overstated. Many factors about a product, such as suitability, quality, and long-term effects, can confound even the most experienced shoppers.[1] It is far from trivial to choose vacation packages, electronic devices, or food produce; for services such as banking, education, and healthcare, buyers' information deficiency is more severe.[2] Yet, in the supply–demand law of economics textbooks, information is conspicuously missing.

When we consider buying a product, the key issue is often how it comes to our attention and how we determine its quality. In fact, consumer markets exhibit fascinating complexities that are often related to information deficiency. Many startups of Silicon Valley exploit a bewildering array of imaginative information strategies that in one way or another impact how consumers find products and how businesses find customers.

Consumer markets play a fundamental role in our economy, and hence Chapters 1–4 are dedicated to them. This chapter discusses a new supply–demand relation, with consumers' limited information as the key feature; Chapters 2 and 3 discuss how information can be improved; and Chapter 4 outlines the consequences of the improvement.

Berkeley economist George Akerlof, in his Nobel Prize-winning work *The Market for Lemons*,[3] explains information asymmetry with a simple example: if buyers cannot determine the quality of used cars there will be few transactions, and market failures can result.

Information deficiency usually occurs in an asymmetrical manner, as vendors know more about the product than do buyers.[4] We generalize Akerlof's insight by postulating that information asymmetry affects most market transactions to varying extents, and only in the worst case does it cause market failures.

Such a gray degree will be called *infocap*—an abbreviation of "consumers' information capabilities."[5] A person with an infinite infocap will understand the product perfectly, another person with zero infocap is totally ignorant about it, and anywhere between, people know, to a degree, the true quality.[6] A higher infocap begets a better understanding of the underlying quality and suitability of a given product or service.

A person's infocap on a given product depends on himself as well as on external factors. His own attributes are skills including experience and talent, and effort comprising diligence given to the task. Consider Gary Kasparov playing speed chess against ten players. He must spread his effort over multiple games simultaneously, and hence his infocap per game is somewhat reduced.

External factors can influence a person's cognitive capabilities. For example, in the early days of Airbnb,[7] the founders improved the visual layout to allow people to better see the listed properties. This simple measure effectively improved its customers' infocap, and the result was a measurable increase of transactions. That infocap can be shifted by external means can be greatly exploited by businesses, as we will discuss in Section 1.3.

An individual can be an expert on a few products and ignorant on many others, hence infocap is relational between a person and a product. Quality is not only about a product being well made or not well made; it is also about personal tastes. A smelly cheese may be delicious for one and repulsive for another, hence quality must include suitability. Having limited infocap, consumers must face all these challenges. First they must be aware that a relevant product exists, then understand its quality, and still compare competing offers.

Infocap is inhomogeneous across the population. If there are two groups with distinct infocap levels, vendors can target them separately. For example, compare a shop in a residential area to another near a railway station.[8] The residential shop's customers tend to be local people who visit it repeatedly, while the railway station shop's customers are mostly travelers who shop there probably only once. Local people usually have a higher infocap level than do travelers. If one shop caters for 80% locals and 20% travelers (80/20), and the other to the inverse proportion (20/80), the shoppers in the first shop will have a higher average infocap than will those in the second shop. Quality:price ratios are usually better in the shops catering mainly for locals than in those for travelers.[9]

The average infocap level can be considered as a public good, because businesses typically target a large fraction of the population: if it stays at a high level, a careless consumer or an out-of-towner may still expect a reasonable deal. It is the diligent consumers' selection pressure that keeps vendors on their toes. Imagine a rich consumer shopping for high-quality goods, with the price as the only cue for quality; the customer would always buy the highest-priced items. He may indeed find rather high-quality products simply by free-riding on the work of others. The collective effort of the consumers exerts an average selection pressure on the vendor, and it is to the average infocap level that it offers its quality/price combination. Since price and quality are not tightly correlated, our lazy but wealthy consumer will obtain quality:price ratios inferior to that of the diligent and experienced shoppers. One can always argue that he is compensated by the free time he might gain.

This section aims to convey the message that consumers know products only up to a degree, which may change depending on many factors to be discussed in detail in this book. The next question is how businesses adjust prices and quality given the average infocap level of their customers.

1.2 Magic pie

The supply–demand law in mainstream economics textbooks, which has two parameters (price and quantity), states that consumers would buy more if price were to drop, and vice versa. However, it implicitly assumes that buyers know quality. In the new supply–demand model we explicitly include quality as an additional parameter. Since quality is only partially understood, it must be accompanied by infocap—the relational parameter between the buyer and the product. The extended supply–demand relation with the two additional parameters is slightly more complex than the mainstream version, but it can account for many information practices of the consumer markets in the real world.

In consumer markets, businesses can change not only prices but also product's quality—sometimes as easily. This is especially true for modern businesses where fast market feedback and lean production permit easy product revision. The mainstream law effectively assumes quality and information as the given constraints, and over the past century it has never relaxed the simplifying assumptions.[10] Once these constraints can shift, interesting consequences ensue.

In this section we outline the new supply–demand relation, which leads to a new theory of markets that we shall describe in Chapters 1–4. (The interested reader may consult the original research work for details.[11]) Consumers with a given infocap level have a buy probability depending on quality and price of the product under consideration. With a higher infocap, the buy probability will be more sensitive to quality variations, demonstrating that the more informed consumer would be likely to buy a product with a more favorable price–quality combination.

The mainstream law states that consumers would buy more if the product were priced cheaper, and the extended supply–demand model supplements this well-known feature with the additional feature that they would also buy more if they know the product better.

To see why consumers would buy more if better informed, consider the following scenarios. If consumers are totally ignorant about the product's quality—that is, infocap being null—they will not buy it; this corresponds to the Akerlof limit of market failures. But if their infocap increases a little, the consumers understand it slightly better, and more transactions may materialize. In other words, the tendency of better-informed consumers would lead to more transactions is an extrapolation from the Akerlof limit.

Transactions between a producer/vendor (hereafter vendor for simplicity) and its customers are usually beneficial to both sides, though seldom equally. The combined benefits can be regarded as an economic pie,[12] with its size depending on the consumers' infocap and the corresponding actions by the vendor. In general, a larger infocap leads to a larger pie, hence we can speak of the magic pie. Resulting from detailed analysis, and assuming that the vendor maximizes its profit given the consumers' infocap constraint, we have the three typical pie charts shown in the Figure.

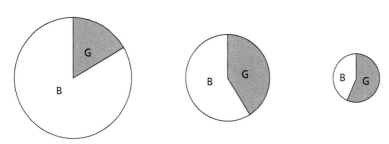

Pie charts and the division of benefits between the vendor (G) and its consumers (B), at three levels of infocap. In the rightmost chart, infocap low, to maximize its profit the vendor would take a bigger percentage of the pie (by adjusting the price and quality), but the absolute size of his slice is small. In the leftmost chart, infocap high, the vendor is obliged by the informed consumers to reduce its share on the much bigger pie. Somewhere for a middle level of infocap, the vendor's profit is maximized among all infocap levels, albeit the pie is not the largest. These are the results if the vendor maximizes its profit on every given infocap level.

How can the additional surplus be generated, as if it is plucked out of thin air? Our theory asserts that economic transactions need not be zero-sum games, that businesses have much more potential to offer than meets the eye, and that they only make an extra effort when they are more pressed by better-informed consumers.

Now let us allow infocap to move. If the average infocap moves, and assuming that the vendor will always be able to adjust quality and price to maximize its profit for each infocap level, we have a one-to-one correspondence between consumers' infocap and vendor's profits. The pie-charts in the Figure can be reinterpreted as the optimal partitions for three sample infocap levels, and in the next Figure the vendor's revenues are plotted for the whole infocap range. At the leftmost point its total profit goes to zero, corresponding to the limit when the customers are totally ignorant; hence there is neither transaction nor profit.

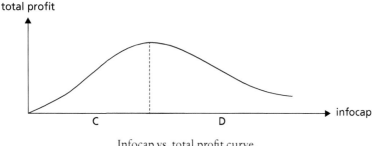

Infocap vs. total profit curve.

Reduced margins are not always bad news for the vendor, as increased sale volumes may still make the profit larger. The total profit increases with infocap until a maximum. The vendor would like his customers' infocap level to stay there; but this is wishful thinking, as it alone cannot keep consumers' infocap fixed.

Why is it that the profit might fall when consumers become "too clever"? While it is easier to understand the rise (C region), the fall (D region) has multiple reasons. We need to consider the consumers' learning curve and competition. First, on mature products consumers become more informed, and their infocap becomes higher. But when the vendor's profit margins are further squeezed, the increase in sale volumes is no longer sufficient to compensate for the diminishing per-unit profit margins.

The main reason for the fall is that the competition becomes more effective with increasing infocap. To show this we need to extend the previous analysis to more than one vendor.[13] The per-unit profits of two competing vendors will, in general, fall. Mainstream economics posits[14] that a profitable niche will attract so many competitors that the margins will be effectively reduced to zero.

This implies that the consumer knows them all. As with the previous example of Kasparov playing speed chess, the consumer must split his infocap among many competing variants and hence must reduce his per-variant infocap.[15] he must weigh the benefits of comparing more variants against the loss of reducing per-variant infocap.[16] Due to limited infocap, adding still more competitors will not reduce the profit margin further.[17] In other words, the competition reduces the margin, but only to an extent. Therefore, limited infocap leads to limited competition, which in turn leads to a finite profit margin.

Increasing infocap will produce two opposing effects: on the one hand it may increase sale volumes, and on the other the more effective competition reduces profit margins; and both effects coexist. We can show that in the C region the former dominates, and in the D region the latter dominates. With increasing infocap the per-unit margin always falls, but in the C region the increased sales volume can offset the decreasing profit margin, which is no longer the case in the D region.

Businesses may not want to fine-tune the quality and price of their products in a search for the maximal profit point. Besides the difficulty of the fine-tuning task, they have other reasons for not pursuing the maximum profit goal at every opportunity.

A firm may voluntarily choose to reduce its profits, and may sell its product at cost or even at a loss. There are economic reasons behind this apparent generosity. A product line may have a long lifetime,[18] and a firm may choose zero or negative profits at the onset in order to rapidly expand its market share. If the firm aims for better cumulated profits over the product lifetime, it is more profitable to not maximize at every time point (hence also at every infocap level).

But do not expect the firm to be always so generous, since in the later stages it may reverse its strategies; in general, we expect it to be initially generous (C region), and to be greedy later (D region), as we will show in Section 1.3.

Companies—especially the fast-growing and innovative ones—must weigh future profits against current profits. Since they operate in a dynamic environment, lowering current profits to grab larger market shares can be more lucrative than optimizing at every point of the lifetime of the product-line. But it is futile to maximize lifetime profits, as there are many strategies to accelerate the penetration of the product that no mathematics can fully account for.

The most important lesson learned from the above market model of the new demand–supply relation, plus the additional considerations of competition, suggests that a business's profit first rising and then falling is a general feature of the consumer market. The rise of infocap is usually accompanied by product maturation, and the profit fall would press businesses to search for either new products or the innovation for more efficient production of the old product.

1.3 Changing infocap

If the infocap constraint were fixed, the previous analysis would lead to another optimization exercise, and this book would end here. However, throughout the rest of this book we shall consider the infocap constraint to be changing, and continue on our journey to explore the reasons and the consequences. We shall see that the widespread rise[19] of infocap is the main pathway that converts apparently non-economic factors such as cognitive capabilities to wealth.

Consumers gradually gain knowledge of a new product, so that the previous profit-vs.-infocap curve can be similarly plotted as a profit-vs.-time curve. In this section we will see that infocap can often be deliberately influenced, and that businesses have motivations and means to either accelerate it or slow it down (or even reverse it). Businesses would try to increase consumers' infocap at the early phases of the product and slow its progression or degrade it at later phases, deliberately stretching the period during which greater profits can be gained. This is the main reason why businesses would try to tinker with consumers' infocap.

When consumers' infocap increases, a firm's per-unit profit margin tends to fall, but its total profit may increase or decrease, depending on in which region infocap is placed in the plot in the second Figure. The firm's attitudes are diagonally opposite: in the *cooperative* region (C) it would help to improve the infocap, while in the *defensive* region (D) it would resist it or even reverse it. In the cooperative region the firm will help consumers to be better informed, and in the defensive region it will use various gambits to sabotage their infocap.

Better information leads to a larger magic pie, but businesses do not necessarily embrace the larger pie voluntarily. In the defensive region D, a vendor can obtain a larger profit from a smaller pie. Since consumers' infocap also depends on external factors, the vendor's marketing strategies can influence the consumers' infocap.

Businesses in the real world may have contradicting attitudes, as their interests are aligned with, or opposed to, that of consumers in the cooperative or defensive regions respectively. Consider automobile retailers, for example. Some have affirmed that "the best customers are informed customers,"[20] but others have claimed that "car dealers might have to deal with informed customers! That must be illegal!"[21] These conflicting statements might both be right, depending on in which region (C or D) they find themselves.

Whether consumers' cognitive information capabilities can be manipulated depends on who is asked. Mainstream academics and marketing gurus apparently hold opposite views. The former downplay or deny that consumers can be duped, and they posit that consumers know what they want and that whatever they do is best for them.[22] The latter insist that for a fee they can manipulate consumers at will.

In the business book *22 Immutable Laws*, for example, the authors[23] suggest that if a product fails it must be the failure of marketing skills rather than the product itself—as if a brilliant marketer could sell any junk to consumers. Such a position is like that of a political candidate blaming the defeat on a faulty campaign but not on the candidacy. We shall see in the rest of this book that the truth can be somewhere between: limited cognitive capabilities leave ample room for information-tion distortion, but their task is not that easy.

In fact, a large portion of product costs goes into marketing, and advertisements are the most visible way for a firm to influence consumers' infocap. Volumes of academic papers have been written on the economics of advertising, and two functions are recognized: to inform, and to persuade.[24] Another function is rarely discussed: firms may distort the truth. There are many ways to describe a complex product, and firms and their marketing agents often promote a biased view by exaggerating positive factors and omitting negative factors. In developed economies, outright cheating is usually disallowed and is occasionally punished. But between lie and truth there is a gray zone; wilful truth-distortion is ubiquitous, but the key lies in the degree. Regulations leave enough holes for vendors with ingenious strategies to tap into the gray zone for profits.

For a firm, it is not enough to have good products; it must care about how consumers find and understand its products and how to stand out from competition. Since consumers' infocap is limited, businesses must split resources between *appearance* and *substance*, and they are obliged to play unproductive marketing gambits.

Manufactured complexities abound in daily life. For example, discount coupons in many disguises may appear puzzling: why are not products priced straightforwardly cheaper? Why do consumers need to cut a coupon out of a newspaper, memorize a promotion code, or mail in a rebate claim to obtain a better deal? Tourist brochures and free maps are full of coupons for restaurants and shops, and if you care to collect the relevant coupons, the deals can be better. Cutting out a coupon and showing it when asked is a credible way to assert that your infocap is above the average and that you care to use it. Consumers' infocap is inhomogeneous, and businesses use manufactured complexity to filter out careful and informed consumers from others.

Rebates are a common practice in the United States, and artificial hurdles to obtain the products at the advertised prices can be daunting. Vendors hope that we sometimes forget to post the rebate letter, or impose strict formalities that may induce technical errors to disqualify our claims. In order to prevail, consumers must often be very persistent. Vendors typically delegate the rebate program to fulfilment firms that have incentives to tout their low rate of successful rebate claims[25] to please the vendors.

Facing the rise of consumers' infocap, must businesses face the dismal fate of seeing profit margins going ever lower? Some put up a fierce fight, trying to slow it down if their products are in the defensive region, but they face diminishing returns fighting the rising trend of infocap. Others quit the defensive region and explore new niches, and new niches in general fall into the cooperative region where competition is less intense.

We shall see that rising infocap does not translate automatically to profit reduction; the key lies in the speed differentials. It is true that fast infocap changes can reduce a firm's profit margins, especially for the laggards in innovation. For the slow rising trend of infocap that we focus on in this book, innovative firms have sufficient time and many innovative ways of avoiding the dismal fate of diminishing returns, by restarting afresh elsewhere.

The relationship between consumers and firms is neither totally antagonistic, as some alarmists claim, nor harmonious, as marketing people and their ideologues would like us to believe. If credibly obliged, businesses have a lead role in wealth creation. There is, in principle, much potential for win–win outcomes, and it underlies the foundation of the market economy.

1.4 Fundamental asymmetry

Between consumers' demands and businesses' supply there is a crucial asymmetry: a given consumer needs a broad array of products and services, whereas a business retains a much narrower focus. This asymmetry is the basis of our theory, and we shall call it the fundamental asymmetry. In this section we shall dissect its origins as well as its consequences.

Tibor Scitovsky[26] portrays consumers as "generalists" and businesses as "specialists." While people's innate cognitive capabilities may be the same, a firm specializes on a narrow niche—its core product—while a consumer must spread his infocap thinly across many niches. Therefore, the ubiquitous information asymmetry arises between businesses and consumers.[27]

The fundamental asymmetry implies that businesses can expand further in the "vertical" direction (specialization), and consumers may broaden their wants in the "horizontal" direction (diversification). The fundamental asymmetry adds an important element to our new market theory, and we shall outline its implications.

For a given consumer there is a magic pie for each of his many wants.[28] Both directions compete for his attention: horizontally for relevant choices, and vertically for evaluating quality. The information load seems an insurmountable challenge for any individual consumer.

The fundamental asymmetry leads to distinct ways with which businesses and consumers gather information about the other side. While information about businesses is public, consumers' personal preferences are rarely fully exposed.

There are a great number of consumers and also a great number of businesses, and on the surface both sides face a great challenge to identify the other side. Consumers' wants are far more difficult to find than businesses' offers, and each of his many wants can be easily saturated—for many items he probably needs only a few units over an extended period. Businesses are not choosy about who their buyers are; all they need is many of them. For example, between the two scenarios of identifying ten consumers precisely wanting an item and having 100 potential consumers each with only a 20% probability of buying, the latter would be preferable. That is why marketers can be content with the use of rough but massive data.

Therefore, the two sides have different tolerances toward substitutability. For a consumer, if he needs a certain item, say a shirt, he cannot accept a raincoat instead, as each of his wants is personal and time-sensitive and cannot be easily substituted. On the other hand, businesses care little about who the buyers are. In other words, the asymmetry in substitutability implies that the consumer's wrong choice would be a bad buy, whereas any buyer's purchase brings good money for the vender.

The fundamental asymmetry leads to distinct hidden potential for the two sides. For an economy to grow, both the consumers' demands and businesses' offers must expand further, and we shall examine the hidden potentials beyond the existent market transactions.

Consumers' wants broaden in the horizontal direction (diversification), and businesses in the vertical direction (specialization, efficiency, and volumes). Consumers often expand their wants spontaneously, and businesses expand their offers with the utmost attention. Moreover, consumers' wants expand slowly, while businesses can expand much faster if needs be. A business always has hidden resourcefulness, which may manifest via lowering prices, higher volumes, and more novelties. Not every firm has the same depth of resourcefulness, and the more innovative and efficient firms may drive the laggards out by competition.

Businesses can much more readily scale up volumes than consumers can with their wants. Should we all suddenly become Big Mac fanatics and eat daily at McDonald's—like Morgan Spurlock did in the movie *Super Size Me*—the fast food chain would easily ramp up its output by flipping 100 times more burgers in a short time. Although in principle the consumers' wants are also unlimited, they are much less ready to scale at will. For example, it may take a generation or longer to convert rice eaters and tea drinkers in China and India to be burger and Coca Cola addicts.

Although businesses have a much larger hidden potential, they will not spontaneously tap into their resourcefulness unless credibly obliged. The most important lesson of fundamental asymmetry is that helping consumers enhance their selection pressure on businesses is the most productive way to expand the economy.

Throughout this book we shall focus on how consumers' infocap can be improved to press businesses more effectively, and how fundamental asymmetry results in a general trend in business models as well as in economic growth.

1.5 New pies

Let us indulge ourselves for a moment in a pipe-dream. Suppose that the consumers' infocap were to improve without end, and one day the so-called information imperfections eliminated, would not the economy settle into the General Equilibrium Nirvana that the founders of economics once dreamed of?

We shall see later that the information-perfect economy will never materialize, and that any ideal limit is but an illusion. The economy will always bring novelties by which severe information deficiency will again emerge.

Businesses do not passively sit waiting the dismal (from the profit viewpoint) fate of General Equilibrium to fall upon them; instead, some drop old products and begin new products to capture the more profitable periods of the product lifetime. New products may bring genuine novelties to the market that contribute towards economic growth.

Consumers will live forever with information asymmetry, but we shall see that it is not as grim as it might seem. While existing products are better understood as consumers' infocap improves, there is no economic reason to aim for perfection, even when people could do so. The infocap improvement opens new frontiers; the mix of the old and the new is in a perpetual renewal.

The magic pie's size does not merely grow; often, new pies on novelties emerge. In other words, products change not only from old to new, but also from few to many. The new pies tend to make products diversify to cater for diverging tastes, so that enhanced infocap is again spread thinly on a growing number of new products.

Therefore, the aggregate consumers' infocap across the economy may increase, but on any given niche it will remain limited. If infocap improvement goes into the vertical direction it can result in more sale volumes, while if it goes into the horizontal direction, new pies may emerge.[29]

For examples of new pies, consider the author of the science-fiction novel *Down and Out in the Magic Kingdom*, Cory Doctorow, who decided to post its electronic version online for free; and yet the paper edition sold more than his publisher had expected.[30] Let us consider how search engines make money via new pies by displaying advertisements next to the search results generated by a query. Even a small percentage of people clicking through to the "sponsor sites" can result in spectacular profits. Google's search business can be considered as a pie, as in a way, Google gives all the pie to its users and obtains revenue from the incidental new pies.

The realization of new pies requires imagination and elaborate planning. For example, tourist-map distributors may simply sell the maps, or they may decide to give them for free and obtain revenues from many local sponsors. Hundreds of Internet startups try hard to invent in business strategies to create new pies while providing their core product for free.

New pies are not automatically novelties, since many businesses' new offers simply encroach on each other's existent products. But with the heightened pressure forcing businesses to expand into new territories, a fraction of the new pies may turn out to be true novelties that further diversify our wants. In Chapter 4 we shall discuss diversification in detail and relate new pies to the "long tail" phenomenon.

New pies can shape the future, so to speak. They are small, they account little for corporate bottom lines initially, and many of them turn out to be false leads, although some will become the mainstay of tomorrow's economy. Firms can ill afford to be complacent about their current successes, as consumer tastes and fads can change within a very short time.

At the outset, a new opportunity seems often at odds with the prevailing consensus, and new pies allow innovative firms to differentiate themselves from their competitors.

Infocap improvement accelerates the introduction of new pies. Some firms are more apt at exploring new pies, while others are more efficient at exploiting an old niche. New pies are usually opportunities that were not previously recognized but become viable because of improvements in infocap and consumer tastes. Entrepreneurs with outstanding foresight and capabilities tend to embrace consumers' infocap improvements, and are more willing to pursue new pies instead of fighting with their customers and competitors over the existing pie.

Information asymmetry on the new pies is, in general, more severe;[31] deep in the "cooperative region," firms have more incentives to go out of their way to help consumers to understand novelties. Since consumers' infocap on any product increases with time, when a business crosses from the cooperative region into the defensive region it should investigate new pie possibilities.

Much of this chapter hinges on infocap improvement. What *could* happen does not mean *must* happen. Chapter 2 and 3 will discuss the means toward the lofty goals. Improvement in consumers' infocap happens neither by magic nor by exhortation to greater vigilance on the part of consumers. Without a full discussion of the means, the magic pie would be only a pie in the sky.

2

Matchmakers

Chapter 2 discusses the role of information intermediaries (information matchmakers) in helping consumers to improve their infocap. You as a consumer have probably never hired a purchase agent for help, so what intermediaries do we speak of? We show that most of your purchases are mediated by information matchmakers, which are everywhere yet are hardly noticed. For example, we do not consider Facebook as a matchmaker, yet its relatively tiny marketing role is vital to its existence. Collectively, these invisible/ubiquitous matchmakers play the pivotal role in consumer markets. We also show that the most effective information matchmakers are those that curate users' own contribution to evaluate products and services. We analyze revenue models and innovative ways to capture consumers' attention and their data in order to play the matchmaker role.

2.1 From middlemen to matchmakers

In Chapter I, much hope is pinned on infocap improvement. We now discuss how infocap can be boosted, and show that enabling institutions and Information Technology (IT) together can make consumers effectively better informed.

Consumers and businesses often find each other via third parties in many disguises that we call "information matchmakers." Their strategic position of knowing both sides confers an informational advantage that neither the vendors nor consumers have.

Traditional middlemen, such as supermarkets, may also be regarded as matchmakers between businesses and consumers, and since these middlemen must hold goods, and whatever they hold would receive preferential treatment, they cannot be expected to provide unbiased information to consumers.[1]

Often, the matchmaking role is not obvious; they are incidental matchmakers. The popular travel guide *Let's Go* has a side role in matching travelers to restaurants and hotels. For example, the guide has a section on Capri—a picturesque island off Naples, Italy—and in the page on "Where to Eat" you may find the restaurants offering the holder of the guide a discount of 10–15%. Those brandishing a copy of *Let's Go* show they have higher infocap than others and care to exercise it. The overall deal can be a win–win–win outcome: the restaurants win additional customers, the tourists receive a better deal, and the publisher of the guide gains loyalty among its readers.

While travel guides perform a matchmaking role on the side, consider professional restaurant guides that mediate between diners and restaurants by publishing evaluations and reviews. Ruth Reichl, a *New York Times* food critic from 1993 to 2001, was regarded as one of the most influential arbiters on the New York City restaurant scene, and restaurants—even the most snobbish ones—were very attentive to her impromptu visits. The matchmaking role played by her (or rather her employer, the *New York Times*), is to leverage economies of scale of one person's expertise for thousands of diners. *Zagat Survey* plays a similar matchmaking role with its coverage of restaurants, but instead of the service of a star food critic it relies on the feedback of many diners.

Big music-companies often use mixed models. They mediate between artists and music fans, so their role may seem to be that of a matchmaker. However, they are commonly known for selling content on physical carriers—discs, for example. Matchmakers can also be governmental and non-governmental institutions whose role is to safeguard deals. Consider *Consumer Reports*, *Better Business Bureau*, ombudsmen, courts, and so on. They aim to keep transactions relatively honest by exposing the worst offenders. Some institutions also aim to protect businesses against delinquent consumers, such as credit bureaus.

Traditional middlemen usually bear the costs of physical transport and storage. Such costs are different from that of evaluation, dissemination, and marketing. For example, Amazon.com uses a hybrid matchmaking model, mailing out books and other products from its own warehouses and also directing orders to myriad of other vendors.

Pure information matchmakers are easy to recognize. For example, stock brokers, travel agents, or real-estate agents, matching buyers and sellers without holding goods. Most of them are Internet services: for example, eBay.com, the auction website. Google can also be regarded as a matchmaker, though we seldom notice its matchmaking role besides search. Its immense economic influence stems from its relatively small for-profit activities, mediating between vendors and consumers through sponsored advertisements. Its users do the matching themselves, vendors bid the relevant keywords, and searchers querying these keywords are likely to succumb to the advertisements.

Why are third parties like matchmakers needed? Matchmakers can effectively improve consumer infocap so that the magic pie can grow, and net surpluses can be created.[2] The more elaborate three-way relationship of buyer–matchmaker–vendor can give rise to win–win–win outcomes. This relationship is ubiquitous in the modern economy under many guises, and it is already the prevailing business model[3] over the dual model of buyer–vendor.

In the last decade of the twentieth century, many Internet services emerged, aiming to be matchmakers between businesses and consumers.[4] When the initial exuberance subsided, it was possible to see that not all dotcoms were a fad. Some popular matchmakers, such as Amazon, Craigslist, eBay, and Priceline, have thrived as an indispensable part of the economy, and many new stars, such as Uber and Airbnb, have also been established. Information matchmakers are not limited to mediating commercial transactions, however. Wikipedia, for example, can be regarded as a matchmaker between providers and seekers of knowledge.

2.2 How can matchmakers help consumers?

Both consumers and businesses face the growing challenge of how to find each other. In this section we focus on how matchmakers help consumers evaluate products in both vertical and horizontal directions. In the vertical direction a consumer already knows what he wants to buy but is not sure of its quality and suitability, while in the horizontal direction a matchmaker may broaden his choices. (Methods by which businesses can find information on consumers' wants will be discussed in Section 2.5).

One of the advantages of a matchmaker is economies of scale. One expert, such as Ruth Reichl, can help many diners; reviews on CNET. com may empower still more prospective buyers. A few specialists can help many consumers find better deals by effectively boosting their infocap. Experts' knowledge can spread without bounds, and the resulting economies of scale hold huge promise for infocap improvement.[5] Leveraging its global reach, an online matchmaker can help people search both wide and deep. "Width" implies a great horizontal range beyond an individual's reach, whereas "depth" means the greater scrutiny of product quality.

A new crop of online matchmakers—let us call them "information platform matchmakers"[6]—have developed a different strategy. Instead of employing experts, they take a backseat as the curator of the consumers' reviews. Such a matchmaker implicitly assumes that other people's experience is useful for you. In fact, this assumption is valid only to an extent depending on the complexity of the subject at hand. Simple matters, such as the cleanness of hotel rooms or the friendliness of restaurant staff, have a high communality. However, the question of whether a holiday package is suitable for you is more personal, and more deliberation is necessary to determine which of other people's opinions may relate to you.

An average person's review is often casual, but it is more intimate and allows people to find their taste-mates. For example, TripAdvisor. com lets its users evaluate hotels and restaurants around the globe. Travelers not only rate businesses with stars but also write detailed stories, allowing others to see for themselves whether the experiences echo with their own.[7]

Reputation plays an important role in markets, but it is an effort-saving substitute for infocap. Suppose a consumer has unlimited infocap (*homo economicus*), knowing everything with infallible clairvoyance. Reputation is superfluous for him: why rely on others if he is all-capable? Everyone else needs reputation as a convenient proxy for infocap. Reputation is built on the experience of others and can be used effortlessly. However, since it is a sweeping judgment we often need to combine it with our own infocap. Indeed, if none of us have ever used our own infocap, there would not be reputation in the first place.

Matchmakers' role as reputation managers is not recognized by everyone. For example, in 2001, Berkeley economist Hal Varian (now Google's Chief Economist) attributed eBay's success to its global reach and suggested that its role was not much different from that of classified advertisements in Sunday newspapers except for the huge size— ignoring eBay's chief role as a reputation manager.[8] Now, many people understand why millions of buyers can confidently buy from strangers, and that the fundamental value of eBay lies in its reputation repertoire. Members with long, impeccable records are more trusted than new members. Paul Resnick and his colleagues[9] found that eBay sellers with long, good records can fetch higher prices for the same products, while Michael Luca found that "a one-star increase in Yelp rating leads to a 5% to 9% increase in revenue."[10]

Uber is a matchmaker connecting passengers and drivers. On the surface it can be regarded as an online extension of traditional call centers, but it can do much more than dispatching call centers: it makes both passengers and drivers more accountable (reputation at stake), and this accountability makes riding on private cars a viable business.

Businesses are rarely interested in one-off deals, as reputation is more important for them than for occasional eBay sellers. Online businesses care more about their reputation than do their brick-and-mortar counterparts, and negative online reviews can warn off many potential customers. Reputation for online vendors is their lifeline, so to speak.[11] If you receive a bad deal from a Sunday newspaper advertisement there is not much you can do easily to have your complaints heard by other prospective consumers. Sunday newspaper readers are isolated, but eBay can leverage collective feedbacks to evaluate a deal.

Firms that stand out from competition may embrace matchmakers' services, as enhanced infocap makes their products easier to shine. If your product is above average, your business can benefit from transparent reviews, even though some are less flattering than your paid PR releases.

One day there will be a widely accepted standard such that any self-respecting business will adopt a recognized feedback-reputation system. There is no need for legislation mandating it, as customers exert sufficient selective pressure for businesses to comply, although institutions (such as government agencies) still have a crucial role to play in certifying such systems. Businesses can proactively embrace such services, for example, by setting up an honest feedback forum where customers can socialize and voice their concerns. Enhanced infocap is a like double-edged sword: the "defensive" gambits would be less effective, but empowered customers' feedback would shed light on what they really appreciate or dislike.

Some businesses may resist being evaluated, but they cannot escape the long reach of reputation. Some matchmakers allow users to evaluate reluctant businesses. For example, Ratemyteachers.com is a popular site on which school pupils and their parents can evaluate teachers.[12] In North America most schools are covered, and teachers receive votes and comments unsolicited. The site remains controversial, however, and is often met with resentment by the schools. Traditionally, teachers evaluate pupils with undisputed authority; on Ratemyteachers.com, students anonymously rate their teachers' competence and demeanor. Questionabledoctors.com allows patients to rate medical doctors.[13] Unless very serious incidents happen, such as when a patient is disabled or dies due to demonstrable malpractice, doctors are quite immune from public scrutiny. Occasionally, insiders speak out and we obtain some rare glimpses of the medical profession, in which both heros and villains are cloaked in white coats and are hard to distinguish.[14]

How far should we pursue accountability? What about privacy issues? In general, individuals having a public role should be subject to the scrutiny, lest they abuse their asymmetrical power. Reputation can be also very personal: we may like and despise people, but we value our own privacy and are wary of reputation dossiers. Ikarma.com allows people to rate their customers and business partners. While few people object to reputation websites for teachers, doctors, and police, we may frown at websites such as Ratemyneighbors.com, Ratemydates.com, Ratemycolleagues.com—and perhaps Ratemyexspouses.com (fear not, thankfully non-existent). The reason is that ordinary folks are about equal in power, and privacy issues outweigh other considerations.

Reputation managers, as a special species of matchmaker, have many tasks ahead. Imagine that one day the records of all public statements are accessible with a few clicks.[15] Weather forecasters must show their track records before they predict for the next few days; finance gurus should disclose their past records before announcing next year's stock trends. Today, such accountability can be obtained only through laborious investigations that only a few determined individuals (such as detectives or investigative journalists) can carry out. With universal accountability coverage, enhanced infocap will lead to much greater selection pressure on practitioners than is currently seen.[16]

Platform matchmakers are ultra-monopolies. The advantage they enjoy is greater than the economies of scale of ordinary monopolies, as it is based on economies of network-scale.[17]

Amazon.com, eBay, and Wikipedia have few rivals of similar size. If you are ranked as the 100th chef in the country you can do a great restaurant business, but if you run an auction website or online bookstore that is ranked 10th in the world, your business may be worthless. Competition between platform matchmakers is particularly fierce, and tends to be a winner-takes-all game.[18]

Among traditional businesses such as bakers and shoemakers there are many to choose from and competition works. In general, the more complex the business model, the larger the pool of alternatives is necessary to select the best. Compared to bakers and shoemakers, online matchmakers are much more complex, with a vast range of possibilities for rules, revenue models, corporate vision, and so on. And yet the number of alternatives is much smaller, which poses a serious problem for competition to be effective.

The huge size of an online matchmaker can be both a boon and a curse. The upside is the enormous economies of network scale, but the downside is that monopolists behave monopolistically by stifling competition and bullying both vendors and buyers with unfair conditions. Given that the stake is higher than ordinary monopolies, matchmakers can go an extra length by playing nice and locking in their first-mover's advantage to reach monopoly fast, and then swinging the monopolistic stick with little worry later.

2.3 Informational division of labor

In the previous section we showed that platform matchmakers can enable consumers to help each other on global scales. In this section we examine further information-sharing and discuss this apparently altruistic behavior, and show that the potential of "people power" can improve consumers' infocap.

Everyone can be an expert in the right place and at the right time. Prominent economists such as Hayek and Arrow have emphasized the importance of individuals' local knowledge for the economy; there is unlimited potential for perceptive matchmakers to channel the right information to the person in need. Informational Division of Labor (IDOL) has an added advantage, as information is often "non-rival:" if I share my knowledge with you, mine is not diminished.

Sharing information among people is as old as human history. In our offline life, the scale of information-sharing is small, and a matchmaker's services are rarely necessary. Through an online matchmaker, however, our input can easily reach millions.

Information matchmakers can rely on experts or on the masses, both of which have strengths and weaknesses. Experts are knowledgeable, but being a minority they can be corrupted more easily, and when they hold the power as the arbiter of consumers' tastes they can be tempted to manipulate the public for their own agendas. Moreover, their services rarely come cheap: to pay Ruth Reichl, *New York Times* must charge consumers more, though indirectly.

Consumers are large in number and hence are less corruptible, as they cannot be bought off in the hope that they will write favorably about your business; and if you did so it would no longer be called bribes but *bona fide* good service. Ordinary folk are often biased, but more by inexperience than by intention. Consumers are in every corner of the economy and represent the richest source of first-hand information. The challenge for a matchmaker is to curate this raw material for insight. Matchmakers good at recommendation can turn personal biases into something useful, identifying taste-mates.

Ruth Reichl kept New York restaurateurs on their toes for a decade. Imagine what millions of diners can do by scrutinizing thousands of restaurants in the city, and indeed all other businesses in the world. Compared to restaurants, services such as education, healthcare, and public offices are far less accountable, and the need of matchmaker-enabled IDOL is more acute. Each consumer would act effectively as an inspector–reviewer, and this would result in powerful selection pressure.

Adam Smith's *The Wealth of Nations* highlights the power of Division of Labor (DOL). Likewise, when IDOL realizes its full power the impact will be no less spectacular. DOL is "limited by the extent of the market," whereas IDOL has no limits in sight. For DOL, workers are substitutable, but for IDOL, people must be treated individually. The IDOL phenomenon has heralded many online business models for harnessing people power.

However, it remains a major challenge to make sense of massive data that are noisy and contradictory, and the solution is far from just a simple matter of patching together all the information together (Chapter 6). We free-ride often and seldom lead: we know something special and better than anybody else only on rare occasions, and generally we are glad to follow the "state of the art." With advanced information-matchmakers extending IDOL further, on each topic a few informed consumers will be able to help an ever larger free-riding majority.

The service of consumers tipping off each other is usually unpaid; consumers are more fellow taste-mates than rivals. The popular adage that information is non-rival needs circumspection, as it is more likely to be valid among consumers than among businesses.[19] People are quick to tell friends or even strangers about good deals or bad traps, whereas businesses are much less inclined to share information about their gullible customers.[20]

Although a swarm of undiscerning diners can spoil a good find, the complacent restaurant may overcharge or downgrade the quality; and yet finding a good restaurant is different from mining gold. The difference is that while gold deposits face diminishing returns—when more is taken, less remains—a business can easily tap into its hidden potential.

In Section 1.4 we asserted that businesses have much more hidden potential to expand, and any item on a consumer's wants list can easily be saturated; if a diner has just had lunch, you cannot force him to eat another one. Businesses can expand vertical offers much more easily than consumers can expand horizontal wants. Whereas it is relatively straightforward for IDOL to evaluate businesses, it is entirely another matter as to how businesses can obtain information about consumer demands (see Sections 2.5 and 3.2). IDOL is a powerful tool that helps consumers to exert selection pressure on businesses unilaterally, because of the fundamental asymmetry.

Consumers' infocap is defined at the individual level, but the aggregate infocap level across the population plays the key role for the market, since business offers are sensitive to it. IDOL can convert individual infocap to aggregate infocap, and below we discuss a special type of IDOL in the recommendation technology.

Recommendation is about "inferred wants." A consumer has much broader wants than he is aware of, and recommendation can expand his wants horizontally. The premise is that in a networked community, one person's opinion may be relevant to others; the aim is to determine the affinity in the "taste space" among people.

Amazon.com began with books: based on your past purchases it hopes that you are more likely to buy recommended books rather than random titles. Many matchmakers try to guess their customers' next wants. For example, Pandora.com and a host of other music sites suggest new songs for music fans, Stumbleupon.com proposes new pages for web surfers, and Google's matching of sponsored advertisements to search queries can be also considered as recommendations. Google's "recommendation precision" is not high—search queries can lead to about 5% click-through rates on sponsored advertisements—but it is still much better than the 0.2% rate for online banner advertisements.[21] Google's relatively better precision (5% vs. 0.2%) makes it one of the most lucrative online businesses. Improving on precision will have the same effect as improving consumers' personal infocap: a larger pie.

In 2006 the video rental company Netflix.com announced that whoever could improve on its recommendation precision by 10% or more would be awarded a prize of $1 million. Thousands of teams around the world took part in the challenge. Based on customers' evaluations of movies, Netflix can "guess" what they may want to watch next. But it is not pursuing a scientific curiosity by teasing the world with a prize, as the improved precision will result in customers enjoying their service more, and the resulting magic pie will become larger.

An additional difficulty is that most wants are not even known by consumers. For every want known explicitly by the consumer there are many implicit wants, yet upon suitable contexts and stimuli, implicit wants may become explicit wants.

We know ourselves better than others know us—or so we thought. Without much fuss about the philosophical implications, IT experts and online businesses exploit the part of knowledge that resides not in our skull but at the crossroads of the community.[22] This in effect violates the implicit and long-cherished doctrine that individuals are the masters of themselves; conceptually, admitting this is nothing short of a revolution!

The potential is huge, as communication data are far easier to analyze and decipher than those hidden among neurons and synapses in our brain. Amazon or Netflix barely scratches the surface of the potential, as they tap only a marginal fraction of information about us. The new science of recommendation already has many applications, and for further advances entrepreneurs' gut feelings alone are no longer sufficient. Multidisciplinary research in big data will bring more powerful algorithms, allowing innovative matchmakers to turn immense potential into real world applications.

2.4 Who pays matchmakers?

Information matchmakers can, in principle, help businesses and consumers. A consumer knows only a tiny fraction of the products and services that might be relevant to him and of which he has only a superficial understanding, while a business and its marketing agents have even less knowledge of their customers. Based on the win–win–win premise, matchmakers are entitled to a slice of the pie.[23] We would expect that who pays more should receive more help, but this turns out to be not the case.

Traditional middlemen have an easier task: they take possession of the goods and profit from price differentials. The difficulty facing information matchmakers is that consumers are reluctant to pay for a matchmaker's services, and this difficulty prevents the matchmaker from entirely aligning itself with them.

This not an affordability issue, but information services suffer an even more severe information asymmetry (Section 6.1). Consumers, being on the weaker side of information asymmetry, cannot distinguish between good service for payment and a downgraded one for free, and they often choose the latter. Here and in the following sections we discuss how matchmakers obviate this dilemma.

What may seem strange is that matchmakers may charge one side and serve the other, and yet this still makes economic sense. Travel websites such as Expedia.com and Hotels.com provide users with tools to search and compare many offers, and are paid by hotels. Buysafe.com certifies the authenticity of goods on eBay for consumers, and is paid by vendors. Similarly, most matchmakers offer mediating services free to consumers, but charge businesses; to avoid the payment difficulty, this is indeed a twisted logic.[24]

Recall that businesses are specialists and consumers are generalists (Section 1.4), although specialists are more calculating than generalists. Relatively careless consumers cannot distinguish a better service (paid) from a downgraded service (free). The message to the matchmaker is blunt: due to their weaker infocap, consumers do not adequately discern quality differences, and hence are less willing to pay for premium services.[25] If the only way of generating more transactions is to entice consumers to participate, the calculating businesses may still agree to bear the costs. This explains the otherwise puzzling phenomenon that penny-pinching businesses pay for a service that may appear to be more beneficial to consumers.[26]

We should not rush into declaring consumers the winners: they obtain the service for free, but ultimately it is they who pay, though this hardly bothers them if indirect payment loops extend beyond the view. The more calculating businesses do not mind, or even prefer, complicated, indirect payment schemes. Consumers are not homogeneous in infocap, and businesses and matchmakers often also propose (paid) premium services for the more discerning consumers. For example, despite many free alternatives in news media, for-payment presses can still flourish.

Yelp.com allows users to search for and evaluate restaurants and other businesses in the United States. When you look for a sushi bar near a neighborhood of San Francisco, for instance, there appears a prominent sponsored offer with a supposedly excellent customer rating. Angie's List, compared with Yelp, goes a step closer towards consumers as it is a user-paid service.

People consider that Craigslist has more monetization power than it exploits. Craigslist and eBay are both dominant players in their respective niches, and their distinct attitudes on how much to charge users show that there is much leeway in business decisions. Less greedy businesses, per the magic pie theory, have better future perspectives. The decision is not only economic, as often entrepreneurs' culture and vision also play a role. Big Music Labels can also be considered matchmakers, but it charges more than do online matchmakers; typically, artists receive only about 8–12% of the total revenue.[27]

This shows that it would be naïve to say that matchmakers offer help wholeheartedly; consumers' cognitive fallibilities require elaborate intermediation, and each matchmaker would exploit their weaknesses as well as help them. Given the large gray zone, there is a plethora of intermediation possibilities; some exploit consumers more, while some help more, depending on entrepreneur vision and the target population.

It is difficult to help "unappreciative" non-paying consumers who are less aware of the subtle difference in quality. Therefore, many MBA graduates help businesses squeezing consumers further for profits; business school manuals and marketing books mostly aim to strengthen the stronger side of information asymmetry, the main reason being that the calculating business side is more willing to pay for their sophisticated services.

And yet the most productive way to grow the economy lies in the other direction, according to fundamental asymmetry. With rapid IT advances, and especially with new business models leveraging these, some innovative entrepreneurs begin to see the great opportunity of genuinely helping the weaker side to promote wealth creation.

2.5 Enticement matchmakers

A consumer's wants are widely spread; for each of his explicit wants there are many more implicit wants. Explicit wants are those on our to-do list, while implicit wants are those that are not in our active memory or that we do not even know we have.

We cannot, or do not bother to, articulate our implicit wants, hence asking consumers to fill out questionnaires would be futile; matchmakers must devise schemes to determine our consumer's implicit wants. For both explicit and implicit wants, information matchmakers' help for consumers is much needed.

We shall show that there are two distinct approaches of exploiting implicit wants: The first type of matchmaker helps businesses to gather intelligence on consumers (discussed later), and the second type of matchmaker is dedicated to consumers.

Why bother with implicit wants when our explicit ones might already overwhelm our to-do list of the day? The short answer is that a surprisingly large fraction of all our purchases is already initiated from our implicit wants. Whether we like it or not, a great number of businesses and their marketing agents constantly scrutinize us for cues of our implicit wants, and many marketing strategies can target advertisements to selectively awaken some of our implicit wants. Implicit wants may play a role as important as, or in time perhaps more important than, explicit wants in the economy.

As discussed previously, businesses' task of finding precise consumer data is difficult. Instead of manipulating consumers using controversial gambits, they often rely on matchmakers with a new strategy: inviting them to play a cool game and watching them surreptitiously for cues of their needs. Since it is so difficult to gather data on consumers, the rationale for the new matchmaking model is to let them self-reveal in the designated places. Often, users cannot help or care that their personal information is exposed while playing.

Let us call them "enticement matchmakers," who hope that many users would engage their enticing services and reveal their wants inadvertently. Most of them are the so-called "invisible matchmakers," which are so shy that most will not even acknowledge their matchmaking role. Enticement matchmakers often start with a useful service, hoping to attract a great number of users, and only later find a money-making model by renting "peeping holes" to the paying businesses.

Hundreds of Internet startups of recent years fit the profile of entice-ment matchmaker in myriad disguises. The *New York Times* reported: "For anyone with a crazy idea for a Web business, the way to make it pay was once obvious: get a lot of visitors and sell ads. Since 2004, venture investors have put $5.1 billion into 828 Web startup companies, and most of them are supported by ads."[28]

How to be paid is the major factor in determining matchmaker busi-ness models; indeed, it underlies the current ecology of the informa-tion matchmaking industry, which is built upon converting selectively our implicit wants to explicit ones. Advertisements are the main path-way to awaken our implicit wants by the current crop of enticement matchmakers; many of the Silicon Valley startups' revenue models are based on the conversion, one way or another.

For illustration, we dissect some well-known services. For many, Google's gmail is an indispensable service; while we use it for daily com-munication, it learns our habits from many intimate facts contained in emails, and may match advertisements to our preferences. Imagine a traditional mailman receiving and dispatching our letters, which he routinely reads for cues of our personal tastes and vulnerabilities on behalf of paying clients. Gmail is such an online mailman, and it uses intimate consumer data to match businesses' offers.

For example, querying Google is an explicit action; the search giant can figure out our related implicit wants and target us with its spon-sored advertisemnts. Similarly, online social networks such as Facebook and Twitter allows members to create trusted relationships and moni-tors their conversations for marketing opportunities.

Advertising agencies aim to convert selectively our implicit wants to explicit wants[29] to suit their profit targets. Tom Starfold, author of *Mind Hacks*, says that "tech companies are basing their entire profit model on the ability to model and manipulate human behavior, but the implica-tion for psychology is, perhaps, more profound."[30]

Enticement matchmakers tend to be generalists, covering a wide range and representing many businesses. McDonald's is only interested in when you would eat the next burger, whereas Expedia's interests are broader but are limited to your travel-related needs. A global entice-ment matchmaker can find, among paying sponsors, almost anything you need.[31] Such a matchmaker covering all the businesses becomes a de facto ultra-generalist, aiming for a large user base and content with low-grade data.

Enticement matchmakers can go far beyond peeking at you while you are on their premises. Facebook, with the recently acquired Atlas tool, can follow a user on computers, smart phones, and other devices if your account is in open status. In a way, a Facebook user is being tracked on all his cyber-activities, far beyond other enticement matchmakers, who only can watch users while using their own services. Many new wearable device makers want to track your offline life by capturing your bodily signals. Do you really want wearable and implantable devices to carry out data mining for corporate profits? In Chapter 3 we discuss a new type of matchmaker that can leverage the technology in full.

3

Personal Assistant

In this chapter we focus on a special type of information matchmakers, totally dedicated to consumers, that we call Personal Assistant (PA). Consistent with our fundamental asymmetry, we argue that the best information matchmakers should not be in the middle between businesses and consumers but should be completely aligned with consumers' interests by cutting revenues from businesses. Although such a PA does not yet exist, we predict that it will be the mainstay matchmaker in the near future. If something is so good, why is it not being utilised? Here it is argued that the initial barriers are high that current matchmaking models first opt for the easier way of obtaining business sponsors, and it is difficult to switch after having become big.

3.1 PA

An emerging business strategy follows the opposite direction of enticement matchmakers, which requires matchmakers to be completely on the side of consumers. Although this is not yet widely recognized, many precursors are already present.

Instead of a myriad of matchmakers sitting somewhere in the middle or siding with businesses, a special matchmaker may focus on serving each individual consumer. Such a dedicated matchmaker should be more appropriately called Personal Assistant (PA). Here we argue that PA may in the future supplant many current matchmakers to be the dominant business model for matching wants with offers.

Many digital assistants are already in use, such as Apple's Siri, Amazon's Alexa, Microsoft's Cortana, Alibaba's TM-genius, and Google Assistant, and they can all take human commands and do many smart things, with the last being probably the best to understand the contexts of our commands. In this chapter we shall explore PA's role in consumer markets—how PA can help us to be smart shoppers and empower us in many aspects in whicht our infocap is deficient.

One may say that these assistants can already carry out online shopping on our behalf, or plan to cover our commercial needs. But in this book, PA differs from them in one crucial aspect: its revenue sources must be cleanly separated from marketing sponsors.

Information Technology (IT) and Artificial Intelligence (AI) place smart PAs on the verge of widespread realization, yet their recent trend faces a daunting barrier. As long as providers are on the payroll of businesses, even the next generation of wearable gadgets that hold great promise for collecting, storing, and analyzing our data would be frowned upon by skeptical consumers. Do we seriously believe that we would embrace all these devices to allow businesses to further analyze us for profit opportunities?

PA is in sharp contrast to enticement matchmakers. Instead of exploiting our vulnerabilities for profits, our PA's loyalty is beyond doubt, and it is given blanket permission to access, capture, and keep all information about us. The implicit wants that PA can capture dwarf those leaked to enticement matchmakers. Our PA monitors our online and offline life far beyond our commercial needs, just as would be a good human personal assistant.

Our PA has limitless memory power at its disposal, and keeps, for instance, all relevant maps, points of interest, preferences, contacts, and past experiences. For some needs, we simply do not know when it is the right time to pursue them. For example, you might have a hobby, and while passing a locality that might trigger your interest without being on your to-do list, such rare, fortuitous occasions would simply be ignored, but with the hugely extended to-do list always in alert status, our PA can choose the most relevant event from many low-frequency events on the long tail, both for your commercial wants and non-commercial needs.

Clearly, as the owner of our PA we cannot do all the potentially interesting things at once. Simple computation, such as being in the vicinity of relevant shops, or your acquaintances, can filter potential suggestions. As the PA becomes more refined, it should be able to determine time-sensitive events, as it knows our calendar and our to-do lists. Whenever we go online, searching, browsing, and socializing, it would observe us in order to produce a more and more complete portrait of our needs. It can access our health data, monitor our diet and sport, and recommend suitable lifestyles.

PA's rise will benefit from the confluence of three trends. The first concerns technology—devices are closer to us and easier to use and hence can capture more data—the second concerns sciences—powerful methods of data analysis — and the third is about economics—a dedicated PA would enjoy the tailwind provided by fundamental asymmetry.

a) Tech trend: unprecedented data collection. From computers to smartphones and on to many wearable devices, information gateways become easier to use and even proactively retrieve data without effort. Easier interfaces allow PAs to capture more of our explicit orders or detect more of our implicit needs, and much can be inferred from our existence without our deliberate actions. Our PA would command a suite of devices and sensors, and be privy to much of our movements, voices, contacts, and surroundings.

   Stretching our imagination a little further, we may foresee a near future in which data from voice, gestures, palpitation, skin moisture, and so on, can add to the unprecedentedly huge amount of data about our existence and interaction with the outside world.

b) Science trend: making sense of growing big-data. Simpler interfaces for us would require much more hard work behind the scenes by refined algorithms that can interpret our ubiquitous inputs and carry out the job in our stead. Recent progress in the sciences of big-data and AI would lend great support to PA.

   AI would focus on the boundaries between our explicit and implicit knowledge, and instead of fearful Sci-Fi robots that would replace and dominate humans, we may envisage a future with a dedicated PA that obeys and empowers us. A new multi-disciplinary science would focus on the interface between man and ubiquitous devices, and encourage progress in the gray zone between implicit and explicit knowledge.

c) Economic trend. PA's loyalty is the deciding factor, and we can already observe the focus shifting to consumers. Fundamental asymmetry (Section 1.4) affirms that the most productive way is to help consumers exert a higher selection pressure on businesses. This is hindered by the practical difficulty that matchmakers cannot easily be paid for their services, hence all the tortuous ways discussed in the previous chapter. Our PA is a single service dedicated to us, and hence it can be paid by us.

   If the payment difficulty can be resolved (Section 3.5), our PA will fully respect the fundamental asymmetry. Of the long list of our implicit commercial wants and social contacts, PA searches for suitable offers according to our priorities.

Before, it was businesses and their agents with sophisticated tools analyzing us, but now it is like turning the flashlight in the opposite direction to scrutinize them. Paradoxically, businesses have much to win as well, as overall economic growth will benefit everyone (Section 3.5).

In the beginning, PA could start with ordinary tasks.[1] Sorting out our massive personal records can be a great help. Some social networks already remind us of birthday greetings that we might want to send, or guests to invite. Our PA knows much more about us than do Facebook or LinkedIn. It may filter information and suggestions from the outside world, it can repeat our unsuccessful queries on search engines with a few variants, and its tasks can be greatly helped by our occasional approvals and corrections via the seamless interfaces.

Today everyone may have dozens or even hundreds of apps on a smartphone, each with a separate function.[2] Among the first steps, PA can act as a super-app for a myriad of apps on our smartphone. We merely give an order for a task, and it is our PA's task to find the relevant function.[3] PA is a command center with a suite of devices, and computing facilities. In the early stage our PA may start from the smartphone and then connect to a few more wearable devices, such as bracelets, rings, and glasses, which act as our PA's external tentacles. As the next step, our PA may retreat to a less visible and secure place, and by then the smartphone will be a front gateway of our PA, and its data may even be stored in commercial clouds.

Paradoxically, one PA platform would probably simultaneously serve millions of people at once, yet in all effects it functions as if it were exclusively for you. There is a synergy for serving many: all of us face the same commercial world, many queries and evaluations are repetitive. and recommender systems can be integrated into PA.

Now we propose a principle that underlies our PA: the principle of connectedness of human needs. This has two postulates. The first states that for an individual, all his explicit and implicit needs are intrinsically connected and should not be treated separately. The second postulate states that everyone's needs are connected, especially those among taste-mates.

The first postulate guides PA looking inwards, as it tries to unearth our needs by collecting, keeping, and analyzing our detailed data, in order to find better matches for us in the consumer markets and in all our interactions with the outside world. The second postulate guides PA looking outwards, cross-checking collective data to be able to achieve an evaluation on virtually every product that can be reviewed, like a good recommendation engine.

3.2 Push vs. pull

To see how PA can help consumers, we first discuss two distinct ways by which people obtain information about products. If a consumer takes the initiative and finds a product it is a "pull;" if a business markets a product to consumers it is a "push." Producers and marketers have been pushing products to consumers for a very long time, but this dominant push model is now being challenged by the nascent pull model, which is enabled in part by progess in IT.[4]

Why is pull better than push? If there were a product variant that all the consumers crave, pull or push would lead to the same end: they would buy it if they could afford it. However, in the real world there are always choices, and each consumer would rank them differently than business rank them. By the pull model a consumer, aided by PA, would begin with his top preferences. On the other hand, by the push model, businesses would prefer to push products according to their own preferences.[5] A consumer's top preferences are, in general, not the top preferences for businesses, and among consumers their preferences also differ from each other.

The pull model may appear to be attractive—and indeed, some have predicted that it would replace the push model. But there is a serious obstacle against this grand goal: often consumers are aoften not able to pull or do not care to pull.

Knowing all the products in the world is not enough for PA to help us. It is more important to know what we really want. Let us divide our wants into two categories: explicit wants and implicit wants. We know when to pursue the former, and the latter are those that we do not know we have.

Of course, PA can act on implicit wants on our behalf, depending on the degree of our authorization. But it can also be an indispensable aid in executing our explicit commands. This is because each explicit task can hide many implicit elements where a good PA might be able to help. The Figure below shows the relationship between explicit wants and implicit wants.

The relationship between explicit wants and implicit wants. (https://www.shutterstock.com/g/Niyazz.)

In this chapter we focus on a special type of matchmaker: ecommerce platforms (or platforms, for short). Let us say that you want to buy a dress, and you go either to Amazon or TMall and type in your relevant parameters, such as size, color, style, price range, and so on. The chances are that there will be hundreds of pages fitting your explicit description. You begin to view the items page after page, and occasionally one item catches your fancy and you may check further details. A discerning shopper can spend hours choosing and comparing the possible items before the final buy.

This online shopping experience tells us two things. First, the number of pages shows that the platform is not sure what you really want. The longer the list, the more ignorant it is about your needs. Second, you cannot articulate your wishes precisely. Out of many choices you finally settle on the final buy, because you recognize a perfect fit when you see one. Hence we speak of many implicit elements in an otherwise explicit task: buying a dress. PA, after spending sufficient time with you, would have learned many details about you and could considerably narrow the range of choices. How can it be so clever? PA has observed your past pleasures and frustrations on buying other items, knows your dressing room and the habits of your clothing, and, moreover, many implicit wants earlier manifested as explicit wants in other contexts. It might even have overheard your conversations about dresses with friends. Most such data you need not states explicitly, as your PA would meticulously take notes about your life by using all the interfaces at its disposal. It is by following your data in its entirety that clever decisions can finally be reached, as if by magic.

Most casual shoppers check far fewer pages than a discerning consumer, and would probably settle on an item among the first few pages, if not on the front page. In fact, most businesses know this, and they pay a lot for ranking high on the platforms.

This shopping scenario, in which shoppers search explicitly for an item, must by our definition be a pull—or so we may think. But in reality, the apparent pull is actually a push in disguise. Businesses pay high fees to be ranked high, and so most shoppers would more probably stumble upon these heavily pushed items on the front pages. Therefore, what we believe is an item being pulled is actually the result of a push. Even for discerning shoppers who dig deep into the long tail, the purchase might still be the combined effect of both pull and push.

Conventional ecommerce platforms boast that they have personalized pages for each consumer, but all they can do is to use your past purchases as cues to guess what you would like to buy next. Moreover, such personalization is more a trick to exploit your weak spots for profits than to genuinely serve you better, which is why consumers are wary in providing detailed feedback.

On the major ecommerce platforms, consumers can, with a few clicks, find the lowest price, but to find the most suitable product is a far more daunting task. Like the previous example of the discerning shopper, spending hours to find the final buy represents a serious bottleneck for the giant platforms who are otherwise proud of their extreme efficiency. This is rooted in the lack of knowledge of your innumerous implicit needs, even when you know you still cannot articulate them clearly to tell the platforms. PA stands to have huge efficiency gains in reducing this bottleneck.

Chapters 2 and 3 concentrated on a product's quality, Q. According to the second postulate, PA can evaluate Q better than can most information matchmakers. But its most significant role is in improving the suitability of products to accommodate our needs. According the first postulate, PA with a deep understanding of all our needs can achieve the best match.

Let us give the benefit of the doubt that the top-ranked item can sometimes genuinely be deemed the best by most consumers after the most severe evaluation. But this item still might not be suited to your personal tastes. If there were a world-best variant for every product category, businesses around the entire world would compete for it (see diversification in Section 4.1). Let us say that the top-ranked product is suitable for one million consumers; its producer would be only too happy to sell it to ten million of them. As a consequence, for the nine million additional consumers who buy the top product, it would be a sub-optimal match for them.

Match quality is different from production quality, as a well-made product can still result in a poor match if it is bought by the wrong person. For this reason we denote match M. Besides Q and M, price p, representing affordability, is also important. Therefore, for commercial purchases the three measures are all important, as together they constitute consumer happiness $H(Q, M, p)$. PA is best poised to help consumers to achieve higher happiness on the markets.

Let us consider Costco as an example. It charges membership fees, and as a matchmaker it is closer to consumers than are most retailers. But it cannot be said to be a proto-PA, as in order to push down prices it reduces varieties. So, we can say that Costco makes good points on p, and to a lesser extent on Q also. There is little Costco does for M.[6]

If PA can help us to realize the pull model, transactions would most probably take place among consumers' top preferences, since producers can readily tap into their hidden potential (Section 1.4); their lower-than-top preferences can still be made profitable. If consumers' top preferences were fulfilled, their satisfaction would lead to more purchases, and the pulled products would also be more personalized and hence more diversified.

3.3 Second-type information asymmetry and privacy

Why is privacy a concern for most people? What are the reasons behind our fears that others, especially businesses and their marketing agents, acquire our personal data? If you ask hundreds of people you would probably receive as many answers. Here we offer the informational explanation as to why consumers should be wary of businesses handling their data.[7]

As an example, consider that you visit an unfamiliar town and are hungry at lunchtime. Nearby, there are ten restaurants that more or less suit you, though you do not know them. The vulnerable information—your being hungry and also new to town—should be handled with care. Suppose the information reaches one of the restaurants. It would push its lunch offer to you, and it may not be necessarily your best match among the ten, as it has probably allocated a larger budget to marketing than to the lunch itself. Being in an unfamiliar place, you might very well take the offer if it is the only one that happens to catch your attention.

Now consider an online platform that knows you and all the ten restaurants. In principle it might propose the best match for you, but it has received different payments, and those who have paid more might be pushed first, Even when all the restaurants paid the same amount, still the platform could give a favor to one particular restaurant for whatever promotional reasons. But the platform might still care for consumers' happiness to an extent; in their algorithms they might strike a compromise between your happiness and the revenues from the businesses. It happens that the platform's suggestion may, on average, be better than a single restaurant's offer, as the former is torn somewhere between C (Consumers) and B (Businesses).

The restaurant owner will not care about fair representation of all the other competitors, as all he wants is to rank high, and if possible to be the highest. But from the platform's view, some balance or fairness is important; after all, the most popular information platforms are often torn between generating good revenues from the businesses and being fair and relevant to the users lest their happiness would degrade them too much. Rarely, the platform would show only one or a few offers, though it is more probable that it would show all of them, but the ranking is impacted by how it is paid by businesses.

Among those ten eateries there would be a ranking from the best to the worst matches (though all are acceptable), if you knew them. Your vulnerable information in the hands of the three (businesses, information platform, and PA) will provide you with three matches differing in happiness.

This scenario shows that there is a second type of information asymmetry hitherto not widely recognized. Until now we have discussed the first type, which is about your ignorance about a given product that you know its existence. The second type is about your ignorance of what alternatives are available for your needs. Moreover, of those that you do know its existence, you do not know them very well. In other words, the first type of information asymmetry is basic, and the second type builds on the first type.

To illustrate their differences, suppose all the items were perfectly known to you; that is, without the first type of information asymmetry, though there can still be the second type because your own needs—especially implicit needs—are not necessarily known. Without knowing them all, it is impossible to establish perfect matches for you, despite the hypothesis that all the products were perfectly known. Therefore, the second type of information asymmetry is about ignorance on both sides (C and B).

Platforms can, to an extent, provide help in overcoming the first type of information asymmetry, but there are two obstacles concerning the second type. The first obstacle is the conflict of interests due to its revenue model, and the second obstacle is its scarce knowledge of consumers' needs, relative to what PA can master. Therefore, platforms' help for consumers is limited.

For example, Google may observe that recently you have frequently searched about a health condition, and based on this it may suggest pertinent information to you, which is surely a good example of serving users better. But at the same time you can be sure that such personal knowledge can help Google to become more efficient in selling its sponsors' advertisements. Therefore, in the gold rush for new data tools, information platforms tend to do good but also be rather dubious.

The final balance might still be in consumers' favor if the platforms exercise enough restraint in their pursuit for profits, and in the absence of PA we may still embrace all the new technologies sponsored by them, though grudgingly. PA, on the other hand, can do much better with the former, as it can know you better, and without any negatives from the latter.

As another example, consider social networks such as Facebook. Its stated purpose of serving people may not be too far from the truth, but if your daily usage doubles from three hours to six hours. For Facebook, this is a great stride in innovation, but your overall well-being, however, is not among Facebook's objectives. In fact, with the doubling of usage time a student's school performance might be adversely affected in giving up much of his offline life, but the company can only be overjoyed with so much more advertisement exposure.

From the informational viewpoint, privacy concerns our unwillingness to expose our weakness to others. In the extreme case of weakness, when in distress we are easy prey for dubious businesses. The recent scandal of a health-products company in China preying on cancer patients has generated wide public denunciation. When we are short on options, the incredibly magic solution appears more appealing than honest doctors' treatments, and desperation makes us lose our usual cool head.

Without the Big-data and AI, and so on, you are just an average consumer and are treated accordingly by businesses.[8] You appreciate anonymity when walking into a restaurant without the owner knowing you are new in town. With new data power, businesses can split us, Big-data and AI can help businesses to identify those who happen to be weaker than average—and we all are weaker in some contexts. Even a more faithful customer of a business can be subjected to price-discrimination, and Amazon and ctrip.com have been reported as price-discriminating against their most faithful (or gullible) customers. Therefore, more data power on the side of businesses can increase profits at the expenses of consumers' happiness.

On the one hand, businesses may want to differentiate us in order to better serve us with personal attention—or so we are told. On the other hand, they can exploit our weak spots for more profits. Like a predator in the wild chasing prey, singling out weaker ones is the standard strategy, and the prey fears to be out of the herd. Hence our ambivalent attitude towards the onslaught of data technologies that businesses are excited about. In fact, additional data power on the side of businesses can be proven to be detrimental to happiness (in the narrow sense of consumer market transactions as measured by the three parameters that compose happiness).

Facing our ignorance of the second type, an individual business would simply profit from it, and would not point out to you that there are other alternatives. The traditional platform would help you to an extent, and might inform you about all the offers, though with some distortion. Rankings are influenced by payments, but this help is certainly better than that of the single business. PA, of course, helps you much more and can preselect a few relevant alternatives from the whole list, according to your needs.

Therefore, for privacy concerns we should be warier of individual businesses being informed of our weakness or ignorance. We may let the traditional platform access some of our data, but without total trust. Only in PA do we have no more concerns about privacy,

3.4 PA's superiority over other matchmakers

In this section we analyze in detail the working mechanism of the PA platform, and show that there are several fundamental advantages over prevailing ecommerce platforms.

Let us again consider a typical purchase experience on a global ecommerce platform such as Alibaba's TMall or Amazon. A consumer would like to buy a pair of sports shoes, but even for a specific query there are hundreds of variants of color, price, and design, and no matter how detailed the specifications, the list is probably very long.

Those products are there because the vendors have probably purchased keywords, as all businesses know that in order to figure prominently on ecommerce platforms they must pay the platforms. For many businesses, online marketing costs are one of most the onerous costs of their products. Typically, the more a vendor spends on marketing, the more likely its products will appear on many buyers' front pages. Good, popular products may appear at the top of the search results, of course, but they might not be the best matches for your needs.

Let us assume that in an advanced phase of PA it can rerank the products according to your real needs. PA can do this better because it knows many details of your habits, usages, activities, and so on, as well as your overall budget constraints. Incidentally, your PA's list will not be too long, as it would show only the short-listed items for you to make your final decision. Whereas the current ecommerce platforms offer acceptable or even good deals to consumers, PA enables much happier matches for you. The difference between the two listings will confer several fundamental, structural advantages on PA over the platforms:

a) Assuming each business can allocate its resources in two categories: marketing vs. product-improvement, or appearance vs. substance. With the advent of the PA platform, investing in the former becomes less effective or downright wasteful, as consumers consider their PA as the unique commercial-information gateway. Therefore, consumers would be insensitive to all the ingenious marketing gambits, as a product's ranking high or low has little relation to whether or not it is found.

On the other hand, any investment in the product itself, either by substantially increasing its quality or by making more suitable products to charter consumers' needs, would bring instant recognition, as the connected PA's (among consumers[9]) make the PA platform powerful in evaluating all products and their subsequent improvements.

Under enhanced selection pressure, each producer has a different resourceful depth to dig in order to produce better and more suitable products. Those who do slightly better would be able to stand out, which obliges mediocre producers to revise or go out of business faster, whereas genuinely good, innovative producers can quickly establish themselves.

b) Happiness. Better products in all three measures (M, Q, and p) make consumers happier, hence their willingness to buy more. Q and p are up to the producer to decide how much substance to put into the product, while M mainly involves the work by the PA platform and other intermediaries (discussed in Section 3.5).

Potential gains for the consumers derive not only from increased investment in product quality, but also from better matches. Improvements of Q and p call for the producer to spend real resources, but improvements of M are purely informational, as there are substantial gains from just matching better.

Paid marketing strategies not only bring poor matches for consumers, but also distort fair representation of product information. In fact, at least in the early phases of PA, it must do much to undo distortion by traditional information platforms.

c) Precision production. PA aims to improve consumer happiness, but incidentally it can also help businesses, and these benefits can be no less spectacular than those which consumers would receive. Moreover, benefiting the two sides can be without the aforementioned conflict of interests if certain arrangements are made.

Here we outline the premises. Until now, the information that producers have about consumers is inconsistent. To an extent, production was based on very rough estimates, with a lot of waste and missed opportunities. Businesses must set aside huge resources for the disposal or under-the-cost sales of unsold wrong products. If a business does a poor job in anticipation, its products often fail, and only occasionally, and with luck, does a product happen to suit many consumers' needs. No matter how much PA can enable consumers' infocap, it is still the producers who first make the products that are to be selected a posteriori. Random production due to weak Business Intelligence (BI) would lead to huge waste, and the failure rates would be also very high among producers. These wastes would be absorbed in the product costs which in turn would be borne by the consumers, and eliminating them would represent a huge boost in overall market efficiency.

In Section 3.5 we shall see that businesses can also make use of massive consumer data PA collets; in fact, PA can guide them during the planning or production stages, and all wastes can, in principle, be avoided.

Therefore, consumers' selection power is the necessary but not sufficient condition for consumer markets, which must be supplemented by BI, requiring businesses to watch for consumers' cues and to revise products accordingly. Consumers' selection works only by elimination of bad products; it cannot play a constructive role in making the desirable products in the first place. Therefore, consumers' selection and businesses' intelligence must work together to make the markets succeed.

Entrepreneurs and firms stuck in the wrong businesses also represent waste on a higher level than the wrong products; as a consequence of PA's new help, businesses can avoid such waste, and can act faster to reorient themselves to other opportunities. The end result is a faster pace in innovation, and ultimately the entire economy would benefit.

The three advantages together would not only improve market efficiency, but also produce victims among the current players. In fact, the marketing industry as we know it would shrink or even disappear, as enticement matchmakers (discussed in Chapter 2), relying on peddling product information, would pale in the shadow of the mighty PA.

Many ingenious marketing strategies by hundreds of Hi-Tech startups would be obliged to abandon their business plans. The third advantage, however, would also lead to a new intermediary industry tailored around PA (discussed in Section 3.5).

Can traditional platforms change their mediating strategies to be the future PA? After all, the giant platforms already have most consumers registered in their databases, and technically they are the most equipped and poised to make PA a reality. Actually, some platforms already place heavy emphasis on shifting attention to consumers, with many innovative strategies, such as Alibaba's New Retail, to better serve consumers and businesses.

It is not yet clear, however, that giant ecommerce platforms can morph themselves into the PA platform outlined here, either by a sudden switch of business plans or by gradually shifting focus from businesses to consumers. Just as a lawyer cannot be paid by both the defendant and the plaintiff, a platform must choose only one side, as serving both of them results in a serious conflict of interests that might present an insurmountable obstacle for a traditional platform transforming itself into PA. It is hard to image giant ecommerce and information platforms foregoing large fees from marketing sponsors and relying on Costco-like membership fees that, in the initial stages, might be small by comparison. In Section 3.5 we shall see that there are many more profit opportunities than are presented by current matchmaking models. PA would herald a new information ecology of consumer markets that can have all the three advantages.

From the PA platform viewpoint (or from any other current information matchmakers), the previous value-creation proposition does not automatically translate into its own profits. Instead of being paid by business sponsors, the PA platform can, in principle, charge for the value surplus that its three advantages would generate.

We shall show that PA may still charge businesses; but ironically, the new revenues could be even greater than the marketing fees that they should renounce, to the point that this could lead to consumers not being charged at all.

Do we return to the same old story that we have just ridiculed? No. It is not that the PA platform reverts to the old ways, but there is a fundamental difference between the new and old revenue models. PA can help businesses in new ways that will not conflict with its mission to be loyal to consumers.

3.5 B1 vs. B2

Until now, we have examined ways of making consumers happy. But what about the happiness of businesses, if any? PA enables consumer happiness by exerting a powerful selection pressure on businesses, as if these were there to be squeezed to yield ever better and cheaper products. Should businesses passively wait to be chosen or eliminated? We shall see that there is much they can do proactively to improve their fate, and help from PA, albeit indirectly, will play a vital role.

The key lies in *when* to help businesses. Consider the process from production to the market. We divide it roughly into two distinct stages: *post-production* (B1) and *prior-production* (B2). Intervening at the earlier stage, would result in huge savings, as discarded ideas are far less costly than discarded products.

With the usual contrast, consumers vs. businesses (C vs. B), C should be further split into C and P. C have commercial wants (products and services that consumers can buy from markets), and P have much more needs, though most of them are not commercial, and a small portion of them can be converted into C's wants. With our economy evolving more and more from material-based towards service-based, we expect more and more P's needs to convert to C's wants.

Similarly, B is now split into B1 (post) and B2 (prior). Therefore, many information intermediaries and platforms must expect the upheaval of having to reposition themselves among P, C, B1, and B2. PA is firmly committed to P, and sits at the top of the information food-chain of consumer markets.[10]

We shall see that although PA would play the paramount role in consumer markets, it alone cannot cover the spectrum spanning from P to B2, as other intermediaries still have room to serve the downstream of the information food-chain. We shall pay special attention to how to help the new position, B2, and emphasize that B1 and B2 should be helped by separate intermediaries.

The data from PA can be sanitized by some services to guide B2 in the product's design, innovation, and evolution, to achieve unprecedented precision–production. But even with much more information about consumers' wants, and even with the best help at the B2 stage, the new products cannot be sure to succeed when they reach the B1 stage. Although the products are now better made and presumably more suitable, failures cannot be avoided completely. However, in comparison with the current situation of most matchmakers helping businesses at the B1 stage by marketing products that are already (relatively blindly) made, savings by precision–production can be spectacular.

P's happiness can be achieved by PA's understanding of their needs; but for B's happiness the businesses must be diligent in adapting and innovating. In comparison with P doing nothing, B's tasks are harder to achieve, as they demand skills and cost resources, and success is not guaranteed.

Product selection is no longer a process of simply choosing it or eliminating it, but rather by a more complex process with both a carrot and a stick. At the B2 stage, businesses receive the carrot (help), but at the B1 stage they meet the stick (selection). Businesses have a greater flexibility to adjust their products than consumers can do with their wants, which justifies our earlier statement that it is more productive to place selection pressure on businesses and not vice versa. This is another manifestation of fundamental asymmetry (Section 1.4).

The rise of the pull model would shift businesses' focus from *how* to produce to *what* to produce; the latter requires anticipation, but the former does not. Pull-facilitating businesses emphasize production flexibility and target smaller groups of consumers. They adapt their products to consumers' needs either by reorientating product lines (disruptively) or ny revising products (incrementally).

The marketing guru Seth Godin, in his popular Seth's Blog, has stated:[11] "Making products for your customers is far more efficient than finding customers for your products." Earlier, Kevin Kelly, in his book *New Rules for New Economy*, wrote to the same effect:[12] "In the coming era, doing the exactly right next thing is far more fruitful than doing the same thing better." Their prescient statements implicitly advocate helping B2 rather than B1. Extending their logic, their statements help us to understand why current matchmakers help at B1 instead of at B2, as "*who* are the target customers?" is an easier question than "*what* do the consumers really want?" The *what* question is deeper than that of the *who* question, since *what* is the ultimate challenge that requires in-depth knowledge about people's needs, for which PA would be best positioned. On the other hand, current matchmakers helping at the B1 stage thrive on the easier task of finding who are the "gullible" consumers distinguished from the average.

The Industrial Revolution considerably enhanced our bodily cap-abilities. Likewise, it is probable that the Information Revolution will extend our mental capabilities—and PA would represent the ultimate embodiment of this revolution. Risks and opportunities abound, and such a perspective should not necessarily frighten us, since humans can remain masters of technology. For those who embrace the next wave in human–technology coevolution, the stakes cannot be higher.

4

Diversification

This chapter asks the following question. With incessant infocap improvements, what would happen to consumer markets? We see that we can never attain perfect knowledge of the products; instead, enhanced selection power leads to product diversification. With the expanding set of consumer products, the expanding infocap, when spread over single items, would just be able to maintain the pace. The proposition that the observed product diversification is mainly due to the average infocap improvements is not widely recognized. The alternative explanation might be that increasing affluence is the cause. The detailed analysis in this chapter marshals copious circumstantial evidence in support of the new hypothesis.

4.1 The principle of diversification

With enhanced selection pressure on businesses, consumer products not only change from the old to the new, but also diversify. Incessant consumer infocap improvement does not lead us closer to perfect understanding but to an endless diversification of product variants.

Infocap improvement will always be met by an increasing number of varieties, so that per-item infocap will always be limited. Consumer infocap on any existent product tends to improve with time; it is the constant introduction of new products that weighs on the aggregate infocap, and the average infocap level (per variant) may appear stagnant. And yet consumer infocap improvement is not in vain, as it sustains the new-replacing-old flow, and powers the few-to-many diversification trend.

Granted that products tend to diversify in consumer markets, who leads the diversification trend? Two opposing views are plausible. One is that consumers have always had an innate predisposition for more varieties but that businesses could not satisfy them before; the other is that businesses always try to impose novelties on consumers with incessant marketing campaigns. We shall see that consumers actually play the lead role.

This assertion may appear puzzling: all diversified products are produced by businesses; in fact, they take a great deal of effort in finding new niches to differentiate their products from each other, sometimes venturing into unproven products and risking their own survival.

First we propose a principle, and then marshal evidence to support it. The diversification principle posits that consumer wants will manifest in all possible ways, unless hindered. People always have a predisposition to have ever broader needs that differ from person to person. The focus will then be on what hinders the consumer's innate diversification propensities. Businesses' natural tendency is to concentrate for efficiency's sake, and though often taking initiatives of diversification, they only do it when they feel obliged to satisfy or even to anticipate consumers' diversifying needs.

Businesses would like to push a product to everyone if they could; consumers, having personal preferences, seldom pull the same variant. The push model aims for efficiency, whereas the pull model promotes diversification. We may attribute the observed diversification trend to the shifting from the prevailing push model to the nascent pull model.

Some businesses (innovative) may embrace consumers' diversifying wants, while some resist the trend. But even the most innovative ones do not embrace diversification wholeheartedly; if they find a novelty with a million suitable consumers, it still prefers to push the product to ten million of them, and they may even succeed by using heavy marketing campaigns.

Businesses may be eager to find new product niches because of competition. Recall that the effectiveness of competition is dependent on consumers' infocap (Section 1.3), hence the empowered consumers are the true drivers. Competition avoidance does not always lead to genuine novelties; we often hear the stories of superficial diversities and novelties without new substance. For example, drugs with the same old active agents may be issued with a new label, but enhanced infocap plus market surveillance institutions may limit such abuses. On the supermarket shelves, hundreds of breakfast brands seem to be distinct varieties, but many of them are similar except for the packaging. Only a fraction of new varieties ends up as true novelties, and they contribute to the diversification trend.

It is remarkable that the diversification trend is an unintentional consequence of markets; individual consumers pursue personal preferences, and they do not care whether collectively their new wants might or might not diversify. Both sides pursue their own interests, inadvertantly resulting in product diversification. Although businesses do it with the utmost attention whereas consumers do it carelessly, it is the latter who are the ultimate drivers.

Consumers' implicit wants (Sections 2.5 and 3.3) also represent their hidden potential and which can be captured and exploited by information matchmakers. Hidden potentials for businesses and consumers are indeed distinct: businesses' hidden potential must be pressed to yield, while consumers' hidden potential should be discovered and cultivated.

Although diversity is desirable for many, some may complain of too much of it, as Barry Schwarz has argued in the book *The Paradox of Choice*. The author showed that shopping has become increasingly complex, and he was frustrated with the bewildering array of variations.[1] If a consumer's infocap is severely limited and his diversity needs are modest, then indeed he may feel overwhelmed. Therefore, selection tasks should be commensurate with selection capabilities.

There are billions of webpages and millions of songs, but no one seriously proposes that there are too many and that we should stop creating new ones. Without search engines we would be completely lost in the labyrinth of the web. When there is a severe mismatch between infocap and many variants, people feel frustrated. In his book *Flow*, Chicago psychologist Mihaly Csíkszentmihályi argues that the right amount of stress, on the verge of being overwhelming, brings the best satisfaction, and he asserts that being either underwhelmed or overwhelmed is not desirable.

John Galbraith, in his book *The Affluent Society*, complains that businesses dictate consumers' tastes; marketing gurus also boast that they can do that. However, in accord with the above principle, there is reason to remain optimistic that eventually consumers' diversifying wants will prevail. Again, it follows from fundamental asymmetry that businesses' efforts to bend consumers' tastes can achieve only limited effects and yield diminishing returns. A century ago, Alfred Marshal said that "the tendency to variation is a chief cause of progress."[2]

4.2 Infocap and diversity

The most significant bottleneck for diversification is limited infocap. Enhanced infocap acts like a magnifying glass so that can see finer details; seeing better is the necessary condition to match diversified wants to diversified products. Therefore, the diversity predisposition is a necessary but not sufficient condition. Facing the increasing number of product variants, a stronger consumers' infocap would make the products diversify, and if it is weakened, the diversification would be reduced.

Eric von Hippel, a professor of MIT Sloan School, observed that on many consumer markets, people do not find what they want but accept a compromise grudgingly.[3] Firms also struggle to determine what consumers really crave. Many consumers and producers not finding each other is not because of the penury that typically befitted former Soviet department stores, but is more like that of a single person who cannot easily find a suitable soul-mate among the millions of people in New York City.

Long tail is a popular concept for product diversity introduced by Chris Anderson,[4] editor of Wired.com. Anderson emphasized that the digital economy boosts product diversification. Erik Brynjolfsson et al.[5] also argued that with progress in IT, searches become more powerful and rare items are found more easily. Many items deep on the long tail are probably on our cognitive fringes, which are particularly sensitive to infocap improvement. In general, most long-tail items represent our low-frequency and implicit needs.

We need to distinguish two levels of diversity: personal and systemic. A single consumer has a personal list of diversified needs, and from one consumer to another their lists tend to differ. The community's diversity measures variations from person to person. Consider the extreme case where everybody's choices happen to be identical: the individual diversity may still be adequate, but the systemic diversity is not.

Internet applications are reputed to be promoting diversity, but detailed scrutiny reveals a mixed picture. For example, recommender systems increase personal diversity but reduce the community's diversity.[6] Repeated uses of a recommender system will deplete person-to-person variations, making their choices similar. Therefore, overusing a recommender system will degrade its effectiveness,[7] which is analogous to overusing antibiotics or chemical fertilizers in other contexts (more use, less effective).

In a strict sense, only the "brave souls" who are the first to stumble upon an untested item can be regarded as truly diversity-enhancing. They inject into the community fresh items to prevent systemic diversity from degenerating. A long tail can become longer due to not only powerful IT tools but also to the people who do not use these; all of us can be such brave souls on occasion. Without IT tools, systemic diversity may be abundant—as in the good old days—but personal diversity remains limited, and with few offline inputs, IT tools would have little on which to work.

IT tools such as search engines and recommender systems can be regarded as diversity converters that deplete systemic diversity to boost personal diversity. Since information is non-rival, such conversion is accompanied by amplification, and hence the diversity on both levels may rise, which gives the impression that IT tools contribute toward making the long tail longer. But with ever more powerful and efficient IT tools, relative weighting of brave souls drops fast, and the next-generation Internet services should take the fate of systemic diversity into account for their long-term sustainability and effectiveness.

While most Internet services seem to help the long tail, it is not obvious how they do it. Consider the social shopping site Groupon.com; suppose it offers a 50% discount at a pizza restaurant in Chicago and a hundred consumers take the bait. Instead of eating at their usual lunch-spots (presumably diversified) they now swarm to it on a given day, and this fact per se is highly diversity-reducing. Over long periods and on national scales, however, Groupon's role is inverted, boosting the long tail. Lesser-known businesses have higher incentives to go for Groupon deals to attract new customers in the hope of retaining some of them as repeat patrons. Since each time Groupon boosts the sales volume of a business from deep down the long tail, by repeating this on many different lesser-known businesses, the thin parts of the long tail are enhanced.

Evaluating items on our explicit to-do list is already hard enough, as implicit wants represent a far larger portion of our needs that rely entirely on matchmakers to mediate, and we shall be either helped by our PA or exploited by other matchmakers. Our PA plays a unique role by keeping a much great number of low-frequency wants in alert status, and on occasion a tiny fraction of them might find unexpected good matches.

4.3 The battle of two waves

In his book *The Third Wave*[8], Toffler observes that our society has been undergoing an epic transition from an industrial age to an information age. Efficiency and diversification waves figuratively battle each other. In other words, the push model (Section 3.2) powers the efficiency wave, and the pull model leads to the diversification wave. If the former wins, businesses would compete for greater efficiency, and the goal would be how to produce. If the latter wins they would instead compete for consumers' attention, and the focus would be what to produce. As per fundamental asymmetry (Section 1.4), those who take their cue from consumers' wishes and capture the precursor of consumer tastes would do better than those who rely on marketing power to push products.

Efficiency from mass production provides the basis of the modern economy. For example, industrialized farming has transformed agriculture. Just a few percent of the US population are farmers, yet they can feed the nation and beyond. Technology in the past century contributed much to the efficiency wave—often at the expense of product diversity.

That product diversity in our economy may seem adequate today is because man's innate diversity propensity is not yet completely suppressed by the efficiency wave. If consumers had not affirmed their diversified wants and were as malleable as John Galbraith once feared,[9] the world would long ago have been dominated by a few standard products. It is not that businesses are shy of trying to impose uniformity for further efficiency gains, but they certainly did, and still do, an impressive job.

Businesses are ambivalent about diversification, and sometimes they anticipate consumers' wants in the hope that informed consumers may appreciate their novelties. Marketing gurus also exhort companies to diversify by telling them to *differentiate or die* (a book title[10]).

Technology trends, business strategies, and institutional policies can all tilt the battle in one or other direction, and yet the long-term trend shows that the diversification wave is unstoppable. Recent indications show that many long tails indeed become longer,[11] and many small merchants collectively gain a larger presence on the Internet.[12]

On the one hand, McDonald's pushes out millions of identical burgers, regardless of local culture and tradition, and organic food on British markets has dropped by 23%.[13] On the other hand, consumers leveraging the pull model can pursue their wants actively by using, for example, the cow-pooling model,[14] which enables them to collectively choose cattle, increasing organic beef sales by 30%. Many indications show that the efficiency wave lags behind the diversification wave. Recently, Amazon.com introduced a new service called Handmade, allowing consumers to find niche products from artisans easier.

The complete pull remains a long-term goal, and matchmakers try hard to aggregate a sufficiently large number of similar demands to channel them to willing businesses. Soon, this "sufficiently large number" may drop when businesses target ever smaller groups of consumers.

Pull is unlikely to win the battle from push overnight. We expect that both will coexist for a long time, and the efficiency wave might even temporarily rollback the diversification wave. For example, Walmart drives down prices by forcing producers into extreme mass production, and local and subtle features are more vulnerable when the price pressure is high. For example, a book exposes the Walmart fish trade[15] in the ruthless pursuit for efficiency, but many factors are not priced in—local corruption and toxic sludge dumped into the Pacific Ocean. In Koeppel's book there is a similar story about bananas.[16]

Some businesses include chemical additives in food, and it might be wondered why they spend money on chemicals with dubious health effects to sabotage their own products. A program on the French M6 TV channel analyzed packaged salads for business lunches. While the naturally fresh look lasted a day or so, the salads remained edible for a week. The vendors therefore introduced several additives—none of them conclusively proven to be harmful—to maintain the fresh appearance for up to a week so that the food was more suitable for large-scale production and retail sale. Businesses in general feel obliged to divert resources from substance to appearance, with costs on both sides: firms spending money on chemicals, and consumers swallowing the chemicals.

Among the happiness elements (M, Q, and p) there is another type of asymmetry. Information about the three elements is much easier for p, harder for Q, and hardest for M. On many comparison shopping websites it is a click away to rank all offers by prices, but it is another matter to determine Q and M. That is why we mentioned that only PA can fully satisfy our M, good ecommerce sites and Costco can satisfy our p and Q, and some businesses go for p only. Information-difficulty asymmetry is detrimental to the long tail and also to consumer markets in the long run, but businesses can profit from exploiting it.

We may have the impression that old businesses promote the efficiency wave and the new, or Internet-enabled businesses promote the diversification wave. Here we want to point out that the new businesses models can do even more damage to the long tail. Comparison shopping websites and ecommerce websites can drive down prices to extreme levels, and quality and suitability (M and Q) suffer as a consequence.

The efficiency wave acquires a new life with progess in IT. Technology can go both ways, and it can also accentuate the asymmetry. One may say that it is the consumers' own fault if they take the bait of very low prices, knowing that the quality is dubious, or prefer appearance over substance—cheaper and better-looking products over more expensive and organic alternatives. In the absence of reliable information, consumers cannot determine the substance well.

No-one should forbid consumers to prefer cheaper products, but most still prefer Q and M, since the if these two elements cannot be satisfied they have no choice but to settle on something cheap. In the long run, businesses with genuinely good Q and M would be forced to match prices or be bankrupted—the classic example of bad money driving out the good money!

Let us turn to another current business model. Chaining many restaurants together into a single brand is a great way toward efficiency, as franchisees (such as McDonald's) follow a company-wide menu, and consumers know exactly what to expect when they step into another outlet thousands of miles away. For example, Olive Garden has hundreds of Italian restaurants under a single brand whose food, in the reviews, is often claimed to be bland.[17] Big chains usually impose uniformity, and subtle features are vulnerable, since rarely any flavor is appreciated universally. For example, garlic is an important ingredient of Italian cuisine, but not everybody likes the smell, and Olive Garden restaurants simply do away with it in most of their dishes.

There is an asymmetry between offending and pleasing features: people react more strongly against offending ones, such as garlic, than the majority complain of its absence. If a business cannot target personal tastes, the only choice left is to avoid controversial features—and indeed the features that no one objects to are small in number, and they result in blandness. Efficiency gains in globalization place severe pressure on many subtle and pleasing features; on the other hand, diversity-boosting tools can help these features. A study[18] shows that wherever the local review site Yelp.com has a strong presence, independent and small restaurants drive out big chains; infocap improvement can fight back the efficiency wave.

Giant firms tend to be good at efficiency, and their edge by clout often outweighs benefits from innovation. They frown on something smallish, but all innovations[19] at first appear hesitant and smallish. They watch smaller firms innovate, and if one of them happens to show promise, they may consider entering. The smaller challengers by necessity must take on risky businesses, and the incumbent giants cannot afford to take similar risks.

For example, big pharmaceuticals are often less innovative than smaller biotech companies. Facing the dilemma of substance vs. appearance, they often choose the latter, as an editorial in *The Economist*[20] claimed in a report entitled: "If the drugs industry is so proud of being innovation-driven, why is it spending twice as much on overhead and marketing as R&D?"

But efficiency and diversification are not necessarily always in conflict. Again, we should avoid stereotypes with sweeping statements that giant firms promote only the efficiency wave. Some big firms can do better by also riding on the diversification wave; for example the Spanish fashion maker Zara takes only two weeks to bring new products from the designer's desk to its shops.

Consider the transition from push to pull in the TV industry. Traditional cable TV pushes content simultaneously to millions of viewers. Internet TV (IPTV) would allow much more interactivity, and people could pull diversified content. For example, a study[21] shows that content consumers in China spend 43% less time watching TV and 45% more time surfing[22] the Internet. Therefore, the pull model increases diversity and the push model reduces it.

Businesses are often torn between the two waves. Some tend to run away from competitors and explore fresher niches, while others like to run towards competitors by being fiercely efficient in exploiting established niches. The former try hard to figure out how to differentiate from the established, while the latter instead prefer to forego the risky new and to chase what is hot. The former can occasionally threaten the latter's domination by disruptively changing business models, and the latter can strangle innovators.[23]

Businesses' anticipation may lead to new products for which consumers may have never explicitly asked. Is this a push? We shall see that such an apparent push is a pull in disguise, in the sense that businesses' anticipation is an integral part of the pull model.

It might seem puzzling that businesses may anticipate the needs that consumers never thought they would have. Great businesses often create the wants we never asked for;[24] novelties may emerge by pure luck, but upon scrutiny it is often the result of hard BI work of the businesses with penetrating insights into consumers' true needs.

The battle between the diversification and efficiency waves can translate into a battle between innovative and competitive firms. The epic trend seems to be clearly one way; fundamental asymmetry (Section 1.4) deems that any victory of the efficiency wave is temporary and that in the end the diversification wave will prevail.

Consumers' predisposition for diversifying wants can be awoken by infocap improvement, and the diversification wave holds the promise to keep pace with the efficiency wave and even to surpass it. Efficiency chasing brings diminishing returns, whereas diversity pursuing yields increasing returns. A half century ago, Toffler predicted that the diversification wave will lead to epic changes in the economy and in society. Many believe that "the future will be personalized,"[25] and that the Age of the Individual that Emile Durkheim[26] once dreamed of might not be too far ahead.

4.4 Apple vs. apples

For the diversification wave, a distinction must be made between two broad classes of product: man-made vs. natural. For the latter we must exercise caution for the long-term sustainability of our unique biosphere.

If in an Apple shop each iPad were slightly different from the others, quality would be suspect. But in a grocery shop, apples identical in shape and color would be less appealing than the irregular ones in home gardens and local farms. For manufactured products such as iPads and cars (Apple class), variations of the same model indicate poor workmanship. As Edward Deming famously quipped, uncontrolled variation is the enemy of quality. If hundreds of Olive Garden restaurants (apples class) propose an identical *spaghetti primavera* dish, many would find it bland—a quality problem of a different kind.

Naturally-grown produce has more complex features, and fruits from home gardens tend to be irregular. It is not that a particular shape or color matches somebody's specific need; rather, it implicitly assures us that the plant has not suffered industrial manipulation. Without interference, fruit and vegetables have a natural tendency to diverge, while the perfection of tomatoes in supermarkets results from chemical treatments.

Man-made objects such as iPads, cars, and even F22 fighter jets, can be considered simple, but apples and bananas are more complex. What are the criteria? Manufactured products are built with explicit purposes, while many biological features are not planned, and we may need them without knowing them all. An F22 has millions of components, and each must be planned for its role and for the overall function. On the other hand, we cannot say that a banana is merely to deliver sugar, minerals, and vitamins, as any single ingredient can be mass produced cheaply. A collection of listed items cannot replace the natural whole, and there is always quality that cannot be quantified. Thankfully, the best food is not (yet) designed by scientists but by chefs, and there are too many features for which a chef's flair is more suitable than a scientist's equations. An untold large number of natural attributes may meet man's many implicit needs, and they contribute to our overall appreciation of a banana or a dish. Likewise, wines are not judged by chemists but by gurus such as Parker. A good meal or wine aims to deliver not only nutrients with chemical labels, but also lesser features that are so numerous that they cannot be counted.

Bananas produced for the intercontinental trade are subject to strict grading; only blemish-free bunches are exported, and modern plantations often rely on the extensive use[27] of pesticides and fungicides. The same crop on millions of acres is particularly vulnerable to insects and diseases, whereas traditional diversified crops on smaller plots are more resistant. In modern agriculture, apple trees are chosen for easier harvesting by machines, and thousands of years of natural selection muct now give way to economic imperatives. In her book *Wealth of Cities*, Jane Jacobs observes that in her native city Toronto, eighteen varieties of apple were reduced to three on the fruit market, because the remaining varieties suit better mass-harvesting and marketing. On modern poultry farms, chickens are uniform in color and size, antibiotics keep them alive in crowded quarters, and constant light makes them eat all the time. For the poultry industry, chickens are a device for protein delivery, but their feathers are of little economic value and are a waste of nutrients. Thus, further efficiency innovations would lead to breeding featherless chickens with shorter feet, wing tips, and beaks (so that the birds would not hurt each other in crowded spaces).[28] One day we may end up with a bird resembling a lump with a food tube passing through it, to achieve still higher efficiency in converting feed to meat.

We may ruthlessly pursue research and innovations in electronics, cars, space technology, and energy with abandon; but scientists in the biosphere must be wary not to inadvertently interfere with Nature and to cause irreparable damages. If an engineers' bridge were to collapse they would learn from the failure and rebuild it, but environmental scientists, bio-engineers, and medical researchers face a higher bar: many environment and biosphere games can be played only once. We need a more reverent attitude in the biosphere than in tech-space. Biologists releasing a new genetically modified apple into Nature should be more circumspective than Apple releasing a new gadget onto the market.

For man-made products, dozens of car makers and a handful of aircraft makers in the world would ensure healthy competition. Yet we may want to preserve all the diversified stock of the 7,000 corns, 1,200 bananas, and hundreds of horse breeds.[29] Suppose the future production of shoes to be dominated by only a few global brands; whenever more diversity is needed, craftsmen would easily pick up their trade again by diving into old archives, but lost biodiversity would be gone forever. Man-made objects can be fully specified and documented, whereas species are seldom restorable. There is only one way to reproduce a species: conserve it carefully. Some grass-root organizations—such as Seed Savers Exchange—now help in the preservation of diversity.[30] Successes in genetic engineering may have side effects of which we are not fully aware, and biodiversity and biosecurity increasingly become many people's concern. In the current economy, our passing needs cannot do justice to the results of millions of years of Natural Selection.[31]

Languages can be also considered to be in the apple class. If languages were strictly for communication, and if the world could communicate in a single language, it would be very efficient; indeed; languages die as fast as one every two weeks.[32] Only recently, people began to take steps to save "inefficient" languages. Man has other values beyond mere efficiency, and marginal languages can enrich our cultural heritages.

Cost–Benefit Analysis (CBA) experts may deride the seemingly extraordinarily high costs of preventing groundwater poisoning as compared to that of preventing highway casualties on a life-saved-per-dollar basis.[33] But highway casualties should be treated differently from chemical casualties, as the latter have far more invisible and long-term consequences than the former. Barren Alaska might have been of little use when the Russian Empire sold it to the United States in the nineteenth century; today, however, its oil and minerals are valuable commodities, and yet many would prefer to keep Alaska National Parks pristine. The industrial need is simple, whereas the diverse values are complex and are much more difficult to price.

Environment impacts cannot be fully measured, yet one must find an acceptable balance between economic development and preservation. There is no optimum, but compromises are needed depending on our current enlightenment level. Our wants are unlimited, but some are more environment-degrading than others, and society should find ways to preferentially channel people to those pastimes that lead to less irreparable damages.

When calling for caution, environment scientists and activists do not necessarily know better than statisticians.[34] They may lose in a debate with CBA experts, as they cannot price insects and plants that stand in the way of resource exploitation with a well-articulated benefits bill.

Whales may avoid extinction due to their gracious appearance, but how many other less lucky creatures are there that have not yet been assigned higher values than for their meat, skin, and bones? The key contention point concerns how much we know of Nature. CBA experts and statisticians are confident that they know all that matters for their calculations. Biologist David Hillis[35] is less optimistic; he has pointed out that less than 10% of all species on Earth are not even described. For the former, often the priceless have no value in their theories; for the latter, precaution must be commensurate with our ignorance with regard to Nature.

The unknown and the less known should not be discarded according to our current economic equations, and we should leave enough pristine Nature for future generations. There will be no magic formula better than CBA or other current yardsticks that can price the present values of the future,[36] and yet precautionary measures must be considered concerning what we can afford, as man's needs and values become ever broader than those of Adam Smith's time.

Our economy can still grow vigorously, but not unbridledly. Efficiency and diversification combine to power economic growth. We should find a compromise between preservation ideals and current economic needs, depending on our culture and values and on society's prevailing consensus which no statistician can compute. We shall return to this issue in Section 9.5.

PART 2

FINANCE AND INFORMATION MARKETS

5

Financial Markets

This chapter extends the consumer market theory to finance, to examine the similarities and particularities of both types of markets. The common feature is still the role of information, and financial markets are also about how to determine a financial product's quality. We propose the market symbiosis hypothesis that portrays the role of financial markets in the general economy. We argue that although financial transactions do not create value per se, all players in the financial markets have a value-creation role in selecting outside investment opportunities. A well-functioning financial market can enable more quality investments to be funded and can stimulate their creation.

5.1 Information and financial markets

Since the good and the mediocre mingle in financial markets, investors, just like consumers, crave quality. The role played by information in consumer markets has many parallels but also many particularities in financial markets.

Are there good investors? Mainstream economics hardly recognizes skills: it teaches us that the market is impossible to beat, and that those who do must have taken irrational risks and have had pure luck. Paul Krugman makes the point in his essay *There'll always be a Soros*, implying that with many reckless speculators, someone will always emerge as lucky as George Soros.[1]

In games that demand skill, chance also plays a role. With poker for example, a hand of good cards certainly helps, but better players can make better use of average cards. In chess tournaments, all moves are open information, and yet many different interpretations and actions ensue. Compared to games, financial markets are much more complex, and skilled players have more opportunies to stand out from competitors.

Just as there is no sure way of becoming a chess or poker champion, it is even harder to prescribe the prerequisites for becoming a top trader.

Understanding the far-reaching implications of facts is the key to win in games and in investment.[2]

Financial markets are intimately connected to the entire economy. News for financial markets is rarely binary, good or bad, and its understanding requires economic, technological, political, social, and historical knowledge. The great challenge for investing is in how to relate disparate information tidbits, to determine their significance a little earlier than others, and to reach actionable conclusions.

Financial markets are all about numbers, and they seem easier to quantify than consumer markets. And yet stock prices often deviate from what is deemed rational by the earnings ratio—sometimes significantly so for a myriad of reasons that numbers alone cannot represent. Idiosyncratic reasons can nonetheless be valid: growth perspectives, patent litigation, a new CEO, a long-term contract in view, and so on. Many Internet companies seem overvalued by standard metrics, and investors also attribute value to perspectives that are hard to quantify. We call innumerous facts financial data, and the complexity overwhelms even the best. A good investor's edge is often intuitive and qualitative, hence it is often ignored by academics well versed only in mathematics.

Let us see how investors select products in financial markets. Retail investors may choose a mutual fund, its managers pick companies to invest, and companies hire employees and executives. In each of the successive stages, some alternatives are selected while others are discarded or reduced. If a startup becomes bigger, early investors (angels or venture capitalists) may offload their stakes to new investors. If the startup fulfils its promises, even these buyers can still make a profit by offloading to later buyers. Unlike consumer products, a company changes while growing, hence its quality is a moving target.

News and sudden events can make immediate splashes in stock prices, and the primary price moves can also generate secondary ripples. A surprise often takes months or years to fully impact financial markets. A bad quarterly earnings report can send stock prices plunging in minutes, but fast price movements may under-react or over-react, and a single event might affect a sector or even the whole market. The subprime crisis of 2008, for example, initially concerned only a few banks, but when the media cried of a recession, a mass psyche set in, ordinary consumers postponed purchases, and the global economy declined with the stock markets.

Experts can be wrong collectively, however. For example, each year *Businessweek* used to ask fifty gurus (mainly chief economists at large investment banks) to predict the DOW index for the following year, but their predictions turned out to be way off the chart. Unlike traders, chief economists may feel more comfortable in being wrong for the "right" reasons than right for the "wrong" reasons. Traders are not burdened with justification, and do not live off theorizing. Chief economists rarely trade, but sell their services by articulating justifications conforming to prevailing theories, and are often schooled in the same way as those who evaluate them.

In financial markets, manufactured fallibility is more rampant than in consumer markets. People have incentives to deliberately distort facts and data, and abuses such as Madoff's fraud scheme are beyond individual investors' due diligence. Moreover, investors will not receive much help from fellow investors and experts, and the motto "information is non-rival" is less true in financial markets than in consumer markets.

Compared to consumer products, financial products face additional difficulties in evaluating quality. First, no-one buys financial products for final consumption, and investors buy them to sell later. Second, whereas for good consumer products one can scale up production easily, but when a financial product is overbought, a good buy can turn into a disappointment. In other words, financial products are less expandable than consumer products.

5.2 Perception and reality

As with dubious marketing practices in consumer markets, we often read that companies manipulate financial data to mislead investors. Why do they try to look better? Are they not afraid of the day of reckoning? Even with imperfect information we may expect that "what goes up must fall down." Confident people would say that the wrong will be righted, implying that the gambits can only cause temporary swings, and that when the dust finally settles the truth will shine.

Strategic gaming of information can bring lasting advantages in financial markets. A company using dubious practices to establish its dominant position early may strangle worthier competitors, and a genuinely good business whose true merits were not recognized in time might be ignored forever. The first mover enjoys many advantages that might compensate its weakness *a posteriori*; the unlucky loser may see its talented employees flee, stock price reduce, and customers shop elsewhere.

Businesses cannot passively wait until truth prevails; they invest in appearance as well as in substance. Some resort to abusive gambits such as creative accounting, but criminal acts may occur less often than the exploitation of legal loopholes in the regulation gray zone. One should not wait confidently until market forces correct the wrong, lest the undeserved head starts creating the *fait accompli* reality.

Perception is more important in the financial markets than in the consumer markets. For physical products, incorrect perception might impact on sales, but they stay unscathed regardless of hyping or bashing. Art is much concerned with perception. An overlooked painting might collect dust in the basement, and yet it can be appreciated even centuries later; but a startup, if not funded in a timely manner, within a short few years will be gone forever. Quality in finance is relative—contingent on many other scenarios—as it is the future perspective that is traded. Physical objects might lay unharmed by the wrong perception and still have a chance later, while investment opportunities can be permanently affected by perception. The financial truth is *perishable!*

Now compare the creditworthiness of a mortgaged house to physical products such as shoes. Along the supply chain, shoes' quality can be inspected by anyone. Financial products can rarely be checked directly; quality often relies on proxies such as past failure statistics. If you pay mortgage bills in a booming economy, banks rarely re-evaluate your default risks. Individual households are harder to check than public companies: retail investors may not visit the company in which they buy stocks, and thanks to analysts' reports and mandated audits they can still have a clearer idea of a household's creditworthiness than can professional lenders. Therefore, information deficiency can be arranged in ascending order: shoes, stocks, and households.

It is easy to understand that stock prices should follow the underlying fundamentals, but it is less frequently acknowledged that prices can also influence fundamentals. In fact, prices and fundamentals influence each other mutually, and a wrong perception can change the reality. George Soros developed his reflexivity theory to show that thinking agents are susceptible to positive and negative feedback loops and hence a science about financial markets should be fundamentally different from natural sciences. People often under-react or over-react, and this generates secondary ripples that in turn feed to financial markets as the new reality. Stock prices do not move around an equilibrium, and price movements and shifting fundamentals are in endless coevolution.

Secondary ripples can be caused by trend-reversing and trend-reinforcing movements, as amply discussed by Soros. Negative forces tend to correct over-reactions, and positive forces are less obvious; price thresholds may trigger stop-loss orders that in turn set off chain reactions, and chain reactions are sometimes due to human cognitive thresholds. For example, the initial signs of the 2008 subprime crisis disturbed little of the mainstay economy, but when Leman Brothers became bankrupt, many took a deeper look. When the Dubai debt default crisis hit in 2009, initially there was no reason for panic by investors afar in, say, Silicon Valley. But those who did suffer losses in Dubai must reduce other holdings in their diversified portfolio. Traders anticipated such fire sales, promptly shorted these assets, and front-ran those who suffered the real losses. Global diversification not only spreads risks; it can also amplify them.

During a financial market crash, billions of dollars of wealth are said to be wiped out, but no cash piles are ever burnt and few of the houses hit by a subprime crisis are demolished. So, what huge losses? Much of the wealth on financial markets is built on future perspectives; during a crisis, people stop taking on risky opportunities. Since capital markets and the real economy are intimately coupled, destruction of perception-based wealth can drag down the real economy.

Suppose that you are the only person in the world who knows a stock's true value. This may seem to be a very good edge, but it would not be of much use if no-one agrees with you, as the majority's action will result in a reality different from that which your original insight represents. There are no absolute fundamentals, but some companies are more resilient in a crisis or have more upside if a boom ensues.

For example, Buffett's value investing strategy aims to identify stocks that are relatively more sound than others, without attempting to predict prices, and the relative advantage can be more useful than the absolute truth. The shifting fundamentals provide profit opportunities for those with capital and trading skills. Soros used to say that a one-eyed king can reign among the blind.[3]

Some theorists write off perception as if the so-called fundamentals are the only deciding factors, while some hold the opposite view that implicitly assumes that in financial markets it is all about perception, such as a beauty contest.[4] In the real markets, both fundamentals and perception influence each other in a hard-to-determine reflexive dance.

5.3 Symbiosis in financial markets

The main role of financial markets is to match capital to investment opportunities, and in principle both borrowers and lenders can be better off (positive sum games). In consumer markets, vendors and consumers can be both winners—hence we speak of the magic pie. Transactions in financial markets are apparently different: one investor's gain is exactly the counterparty's loss. So how are positive sum games possible? The short answer is that there is participation from economic agents outside financial markets who are ultimately responsible for value creation. Complex dealings within financial markets do not create value directly, but their overall role is to select investment opportunities from the outside economy.

There are direct and indirect transactions by outside companies on financial markets. Direct: they may do an IPO and occasional secondary offerings, or issue bonds. Indirect: stock prices can influence their negotiation position, and it is better to poach and keep talents using stock options instead of cash.

Financial markets relate to the rest of the economy, which acts as an inexhaustible value source, and they are open games without fixed players. We may regard the investors closer to the outside economy as upstream, and those further away as downstream. Upstream risks are concentrated, investors there disgorge opportunities downward, and many layers are necessary to digest the raw opportunities with attendant risks. Most investors in financial markets have downstream outlets. Raw opportunities from the outside economy arrive in bursts, and the shocks are gradually absorbed going downstream.

If traders deal only with each other, sooner or later they would realize that they play a zero-sum game. Losers would either quit or be bankrupted, and the game would stop. Sustainable financial markets are different from casinos in a crucial way, in that most players can be winners. Someone must have provided profitable opportunities in the first place, and frequently replenish them.

Who is so altruistic or careless? Why do not the upstream investors keep all the undervalued shares for themselves? The short answer to this paradox is that there is no sure bet; they do not keep all the profit potential, and are obliged to cede a portion for downstream investors. This may seem as if a man picks up a $20 bill and leaves a change of $10 on the street. Due to risks, the proverbial dollar bills on Wall Street may indeed never be finished.

An investment banker might not necessarily be tempted by more profitable windfalls of a venture capitalist, who may not want to build a company himself. Angel investors and venture capitalists are near the upstream opportunities, and incubators of startups are even further upstream. Downstream profit margins tend to be smaller, but none-theless can be lucrative, as big hedge funds can scavenge them and arbitrage the tiny differentials with massive sums. It is not necessarily a case of "higher upstream the better," depending on the investors' skills and accesses. Undigested risky opportunities are offloaded as "waste," but they can be fresh opportunities for investors downstream. Well-functioning intermediate layers may stimulate exploration and expansion upstream.

Similarly, home buyers are another source from the outside economy. For example, instead of buying a house until you save enough, mortgages allow you to access it much earlier in exchange for your future earning power. Banks package mortgage contracts to pass on to downstream layers in the finance food chain.

Financial markets in and of themselves do not create wealth, but the outside economy does. And yet financial markets have a key role to play: agents on each intermediate layer not only pass on profit opportunities and attendant risks downstream; they also select. They do not blindly accept whatever is disgorged from the layer just above; they preferentially choose some and discard others.

Cumulated selection by all the agents of financial markets facilitates financing the wealth-creation opportunities of the outside economy. There is a grand symbiosis between financial markets and the outside economy.

It is convenient to split investors into two types: risk-hedgers and risk-takers. The former tend to be upstream, and the latter downstream. The hedgers are closer to the outside economy, and the takers (usually more diversified) can absorb more risks. Since there are many intermediate layers, an investor can be a hedger with respect to his downstream trading parties, and a taker is in the other direction.

Many professional traders frankly admit that they have scant interest in the fundamentals of the real economy, and are content to rely on trading skills to live off the profit opportunities generated by others. And yet they can still unwittingly play a role in the symbiosis, which might benefit the outside economy indirectly.

Upstream information is idiosyncratic and thus hard to quantify. For example, individual mortgages are idiosyncratic, while repackaged mortgages can be quantified by fancy mathematics tools. Upstream investors (such as angels or VCs) weigh potential windfalls vs. risks more by innate feelings rather than by number-crunching, while downstream investors (quants) deal more with numbers and often rely on computer algorithms. There is a gradual scale ranging from the less quantifiable upstream to the more quantifiable downstream.

Mainstream finance theory focuses on quantifiable data and ignores non-quantifiable information, hence misses the most important link between financial markets and the outside economy. The quantifiable part is easy to teach, but the non-quantifiable part requires experience and broad knowledge that textbooks rarely provide. Since mainstream finance indoctrinates students predominantly with quantitative tools, most of them are attracted to the better-quantified downstream, instead of to the outside economy where the value is created.

Without the symbiosis perspective, it is hard to justify the existence of much of the finance industry. On Wall Street, thousands of financiers sift bits and pieces of information, so what role do they play for the economy besides their own profits? We may understand the people's anger when greedy bankers caused havoc on the world economy. They are variously labeled as gamblers,[5] parasites, or even bandits with animal spirits,[6] but none of them hint at a positive role. Our view is that financial markets can fulfil a vital selection role in the economy, and even greedy financiers who think of nothing else except their own profits can still have a positive side-role, despite themselves.

The key difference between casinos and financial markets is that the latter can on average be a slightly positive sum game. Spontaneous symbiosis can still happen, but this will be less and less likely when financial markets become more complex, and it will gradually be replaced by the conditional symbiosis that must be deliberately cared for. The invisible hand alone is not sure to channel greediness to good causes; the challenge will be for market institutions to create favorable conditions on which further symbiosis can be fostered. Cumulated selection by investors can have lasting impact on the economy. Over time, skilled investors would have more influence with their growing capital, and their strategies might be copied by other investors. Such cumulated effects will contribute to the entire economy.

Mainstream theory often writes off the complexity of financial markets. For example, one finance professor went so far as to assert that a single super trader suffices to keep the prices right for the rest of us.[7] Besides information limitations, this is wrong when market impacts are considered. The saying "information is non-rival" is less true on financial markets than on consumer markets. Market impacts make the free riding strategy ineffective. Even a super trader himself must expect performance degradation as he commits larger capital on a given niche.

For example, successful hedge funds are reported to return cash to investors, and to turn new clients away, as their selection power cannot replicate the same level of returns for larger capitals. The implication is that more people must work and find fresh and diversified strategies. Free-riding is still an acceptable strategy for many who do not care or who are not sufficiently skilled, but they tend to make less than the drivers do.

We must refrain from blindly cheering that the greedy financiers do well by doing good, as only a tiny fraction from their trading contributes toward useful selection work that facilitates symbiosis with the outside economy. Bees, though, only go for the honey, and their inadvertent symbiosis with agriculture can yield benefits far beyond the value of the honey. Suppose that 99% of all trading volume is attributed to traders' gaming each other and to noise trading, and that 1% contributes towards the beneficial effects for the economy. Are financial markets effective? Even future market improvements may not take us anywhere near the effectiveness of bee–agriculture symbiosis. The financial market symbiosis will probably remain limited, but its role in the economy is irreplaceable.

How large should be the finance sector relative to the real economy? While acknowledging the symbiotic role of financial markets with the real economy, the public cannot stop noticing the naked self-interest and outrageous excesses of the financiers. Many people now realize that the conditional symbiosis on financial markets can fail without regulation.

The complexity and sophistication of financial markets make them appear to be over-bloated. People now question whether all the complexities are really needed for the selection role. In financial markets, manufactured complexity abounds. Many complex derivatives arise from exotic innovations designed by big banks for promoting more transactions without worrying about the concomitant systemic risks.

5.4 Diversification vs. concentration

Financial markets need to fulfil two basic functions: to select quality investments, and to diversify risks. Selection is the aim, and diversification is the tool. Investments carry risks beyond anyone's control, and selection alone is not enough, as all investors must spread their risks to an extent.

In a typical transaction the risk-hedgers offload a part of the profit potential to the risk-takers, along with the attendant risks. The layers downstream not only spread the risks but also select upstream risky opportunities; investors pick some and shun some. A retail investor's capital ends up in many upstream companies, and a company sees its shares going to many hands downstream. On this multilayered structure, each node passes on the selection pressure upstream, whereas the risks and opportunities trickle downstream.

An investor must face the dilemma between diversification and concentration, but we shall see that neither extreme is desirable.[8] There is benefit in going after depth as well as after breadth; the two directions compete for investors' limited infocap,[9] similar to consumers' dilemma facing many product variants.

Finance textbooks tout the advantage of diversification, and seldom admit the need for relative concentration. Buffett regarded his investment strategy as deeply understanding a limited number of companies and investing only in those. The balance between breadth and depth also shifts in time; one typical analyst covered seventy-five stocks in 1967 but only twelve in 1998, as quoted by Thomas Friedman in his book *The Lexus and the Olive Tree*.[10] Three decades make the information load per stock heavier.

We do not expect skilled investors such as Buffett or Soros to run index funds, as they have a deeper understanding of some investments. If you know a few sectors and companies, you should overweight these. In general, the more informed you are, the further the deviation from the maximal diversification doctrine—but even the most informed investors still keep some diversification.

If there were only a few parameters to describe a stock, there would be no depth worth checking. Analysts and investors combing minuscule details would simply waste their time; as there would be a theoretically optimal (maximally diversified) portfolio that would beat all other strategies.

Idiosyncratic depth is difficult to quantify. It cannot be expressed in mathematical terms, and hence it is ignored in mainstream finance theory. And yet going after depth can manifest in measurable ways; most mutual funds invest only in a small fraction of all the tradeable stocks. Theorists are silent on this systematic deviation from the maximal-diversification doctrine, and they cannot merely deride the whole mutual fund industry as wrongheaded or fraud.

Therefore, limited information capabilities account for the observed limited diversification. Diversity is the admission of our infocap deficiency; the more informed diversify less, and vice versa. Admitting one's own limit is a virtue not shared by many, and both amateurs and experts can fail. Over-confidence[11] in their own capabilities and failure to diversify adequately are often reported. At the opposite extreme, maximal-diversification doctrine fails to recognize the selection role at all.

There is a twist in our praising selection and bashing maximal-diversification. If an investor knows a few stocks or sectors, as many of us do, should he invest only in these, blindly following Buffett's motto of buy-what-thou-know? He could fare less well than those knowing nothing but buying into an index fund. Let us call this the driver's dilemma. He would contribute to the overall selection effort sustaining the index, yet his own gains tend be less than that of the free-riders who give up driving!

Detailed analysis[12] shows that above a threshold level of infocap you should trust your own knowledge and invest exclusively in what you know. There is no cheer for mainstream finance theory yet: for those below the threshold, a hybrid strategy is the best: overweight on the stocks they know, but still a portion of their wealth in a maximal-diversified portfolio. Even slight knowledge will warrant deviation from the maximal-diversification designed for total ignorance, and the result is that investors with any infocap level can beat the free-riders. We must bear in mind that if all investors have a share in index funds, the index itself would perform less than it does now.

Prior to the subprime crisis of 2008–09, weakened selection allowed mortgage lenders to load up massive junk debts, while the original risks were camouflaged as if diversification could do magic to junk. Lenders could divest[13] entirely from their dubious debts and offload them downstream. Their incentives were wrongly aligned: no more due diligence for selection, and they even used "teaser rates" to lure people to borrow with abandon.

Diversification of risks is premised on the assumption that the good and bad mingle and that diversification can result in average quality. Without selection at all, the bad has a higher chance appearing. For example, incestuous gambits such as teaser rates would rapidly degrade the debt quality.

Diversification brings hidden risks of which finance students are rarely warned; it is less effective when more people use it, and it often fails precisely when it is most needed.[14] Diversification reduces the individual risks but builds up systemic risks. Stocks tend to be particularly correlated in a crisis; for example, during a market crash.[15] Proponents of index funds never mention the consequences of all investors following their advice. Diversification per se is a good thing, all other things being equal; but the caveat is that many other things are not equal but degrade, making the intended maximal diversification less effective or downright harmful.

One special reason for concentration is not of informational origin. Some powerful insiders are not only better informed but can also change the company, and such insiders will have added incentives to be even more concentrated. Indeed, owners and top executives often hold large stakes in their company. Is the over-concentration a sign of love for the company they own? Perhaps. But there are economic reasons behind the commitment.

For most investors, selection means take it or leave it, even for the most informed analyst. The powerful insiders concentrate on their own company, not just for being better informed, but most importantly, if they exert an extra drop of sweat in many invisible ways they can create fundamentals that no outsider, however well informed, can. The insider/outsider line is sometimes blurred. Powerful investors such as Buffett often demand a board seat in a company in which they invest, and VCs often want to have a hand in managing a startup that they fund, to guide its course.

It is seldom mentioned that companies seeking capital need also select quality investors and obtain money from diversified sources. Let us call it "reversed diversification." Unlike the quality of investment opportunities being hard to determine, money is money, so why should companies be picky about its sources?

Besides money, a good VC can bring expertise and prestige to a startup connection. If a financial backer has full understanding of the startup, he may be more supportive in needy times. In international finance, too much money can sometimes be a bad thing. We frequently hear of hot money flooding into a country with little economic justification, and it can cause havoc for its economy. Speculators with large sums and little understanding of the opportunities in which they invest can enter and exit without much regard to the underlying. They usually have short horizons and no commitments. Reversed diversification is analogous to, but different from, that for investors, in that capital seekers appreciate diversified capital sources and want capital from people with understanding and commitment.

5.5 Marginally efficient markets

Granted that there are exploitable opportunities, how effectively can investors arbitrage them away? Physical resources such as gold deposits can be exhausted, but profit opportunities on financial markets will never be so, regardless of the capability of investors. Financial markets can enable symbiosis, and market efficiency measures how well they do.

Efficiency is measured by degree, and some markets are more efficient than others.[16] If profitable trading opportunities are more difficult to find in one market than in another, the former is said to be more efficient than the latter; but perfect efficiency will never be reachable, for important reasons to be explained.

Profitable patterns attract risk-takers, and their exploitation will weaken the patterns; hence the market may become more efficient. Smaller margins require larger trading volumes[17] to exploit, and the perfect efficiency limit would need unlimited capital. More participation of risk-takers, if they do it right, will make investment opportunities fairer valued, and outside businesses may also be more eager to join; and this in turn will replenish the market with new exploitable patterns. On the other hand, when a market becomes more efficient, risk-takers will find it less interesting (diminishing returns), but outside participants will find it more attractive. There are many investors but also many investment opportunities, and total selection power is limited on any instrument. For an investor there are probably many similarly attractive opportunities competing for his attention.

It is impossible to attain perfect efficiency. Whenever the limit is approached, some traders would leave, and new participants would join from the outside economy. Both of these actions prevent us from going closer to the ideal limit. Therefore, financial markets should be said to be *marginally efficient*. The margins adjust the ratio between the risk-takers and the hedgers, and when they become smaller the takers start to leave and the hedgers are more willing to join, and vice versa.

A well-functioning market will be able to keep many risk-takers interested, despite small profit margins. The market will be less efficient when it is tumultuous; if risk-takers are scared, profitable opportunities cannot find buyers, and outside businesses reduce or hold off their financing operations, fire people, or delay IPO.

Low-lying opportunities can be exploited and exhausted fast, but more complex patterns take longer to be understood and are harder to exploit. A study by Doyne Farmer[18] et al. found that the exploitable patterns persisted over twenty-three years. Traders exploiting profit opportunities will leave their own exploitable patterns that in turn feed to other traders. This is especially true when amateur traders join the game, as it is probable that they leave more patterns than they can profit from. On financial market ecology, prey and predator frequently switch sides, and it is not easy to distinguish who has come to eat the lunch and who are the lunch, as winners often hold only slight edges over losers.

Efficiency is a relative concept that depends on being informed vs. uninformed and interested vs. uninterested. For outsiders the market appears efficient, but for experienced traders it is often not so, as prices can be right for the former while wrong for the latter. Experts in one niche can be ignorant in many others, and those skilful at some moments might miss profit/risk signals at other times. But even for the most skilled and successful traders the markets still often appear efficient. Perfect efficiency would imply that no-one could find any exploitable opportunities over extended periods.

On-looking participants may join the markets, and existing participants may quit. The reserve pool is not accounted for in any statistics, yet the onlookers may play just as important a role as those already in the game. It is like manipulating the visible part of an iceberg. Even if that is the only part of interest, the larger submerged part must be considered; changes on one part beget the other part reacting.[19]

There are indeed games outside the current game, and there is no point in speaking of isolated efficiency without considering impacts on the whole economy. Reflexive impacts between participants and onlookers in financial markets can continue without end, as evidenced in the following example.

In the economics literature, a well-known puzzle concerns equity risk premium (ERP). Over a century, stocks consistently outperformed risk-free bonds by about 7%.[20] This baffles economists, as investors needed to be exceedingly risk-averse for not having arbitraged away the margins over such a long time. Based on mainstream equilibrium methodology, many proffer solutions to this puzzle—and years later it remains a puzzle.

In this marginally efficient market (MEM) theory, there is a new way out of the puzzle. Take, for example, Dow30 during the past century; of the original components, only a handful remain in the current index. The combined capitalization is thousands of times larger today, and most of the components of the index have been replaced. While more and more people chase profit margins, more and more new businesses list stocks in the financial markets and hence replenish trading opportunities.

Consider the non-equilibrium metaphor: a river. It would be wrong to use the theory describing a pond to explain a river. Although a river can be almost as flat as a pond, it is an open system with sources and outlets; likewise, it would be wrong to force a theory of general equilibrium upon the financial markets, which are also an open system. The previous mainstream ERP solutions do not consider sources from the outside economy, and dynamics cannot be accommodated in equilibrium theories.

One of the central tenets of modern finance theory is the efficient market hypothesis (EMH), which proves more irrelevant than wrong. As Buffett has expressed it, adherents of this theory "observe correctly that the market was frequently efficient and went on to conclude incorrectly that it was always efficient."[21]

Our marginally efficient market theory bypasses the academic debate as to whether markets are efficient or not. Some see inefficiencies and most do not, all for valid reasons. The minority who do see and exploit inefficiencies perform the selection work and those who do not, free ride.

5.6 Financial institutions

Efficiency can be defined on different levels. On the lower level, financiers are already doing their best in pursuit of profits. On the higher level, the stakes are also higher: institutional rules can improve systemic selection, just like matchmakers can make consumers collectively more clever. Improvements on the higher level have higher potential which clever traders on their own cannot attain. Efficiency on a lower level is measured against an optimized benchmark under constraints, but efficiency on the higher level considers structural changes, and the possibility space is so huge that no optimization can be contemplated.

Traders win or lose, and market forces take care of rewarding or punishing them. This is no longer true on the higher level, where wins and losses go to the whole economy. It is hard to reward or punish on the higher level; when a wrong policy caused crises with trillions of dollars of losses, few culprits were ever punished. The higher institutions are a public good. They hold the key for economic development; but are often hijacked by powerful insiders for their own agendas.

The notion of the higher level is not to be confounded with higher capabilities. How can a regulator know better than the traders about the correct price or when the next crash will happen?[22] On the higher level, the bigger picture allows regulators to monitor macro signs of imbalance, and preventive measures are possible.

Institutional designs matter greatly for the grand market symbiosis, and the lack of them, or wrong designs, can degrade selection on financial markets. For example, investment research provided by analysts is often paid for by the companies they cover, so how can we expect information to be objective? It is like food inspectors being paid by the fast-food industry, or lung health reports financed by the tobacco industry.

When mortgage lenders divested entirely from the risks, the floodgates were opened to bad loans that turned out to be a time-bomb for the whole economy. Rating agencies also have an incestuous relationship with banks, and whether or not they pay a fee can make a big difference in a bank's credit standing. Imagine a legal system in which defendants can hire not only lawyers but also judges to deal with their case; the government keeps judges and police on its payroll for a purpose.

In a global crisis the inefficiency of the old age actually saved some banks. Giulio Tremonti—at that time the Italian finance minister—once remarked that Italian banks were relatively unscathed in the 2008 subprime crisis, as the bankers were simply less fluent in English and were left out of the last wave of financial innovations. Their simpler, older systems are less prone to the systemic risks. Erste Bank in Germany also has its CEO to thank, as he deliberately wanted to keep banking simple in order to avoid being affected.[23]

Some recent innovations might instead enhance informational selection on financial markets. For example, the peer-to-peer lending and borrowing services Prosper in the Unites States and Zopa in Europe, leveraging social trust relationships of borrowers and lenders. It may be objected that amateur credit checking may fare worse than bankers, but supporters say that intimate personal contact provides a more detailed understanding of default risk than does software used by big banks.

These platforms are matchmakers between lenders and borrowers, allowing people to assist themselves in enhancing their collective information capabilities. For funding the startup businesses there are Angels and Venture Capitals. At one time, Angel investing was for an exclusive club of the wealthy, but the platform *Angel-list* now makes access easier, and upstream selection more effective. Enhanced efficiency again attracts more startup candidates to line up to seek funding, boosting the symbiosis mentioned previously.

Distinguishing the two levels is important in identifying the causes of a crisis and in prescribing remedies. During the subprime crisis, Alan Greenspan (former US Fed chairman) was shocked to learn that the bankers were very greedy, hence conveniently laying the blame on them.[24] Since Adam Smith, we have not faulted self-interest alone when something has gone wrong. Making money can cause negative sum games and positive sum games in the economy. Suppose that a zoo keeper were to inadvertently let a tiger out of its cage, and then be shocked to learn that it had killed bystanders. Tigers express animal spirits when they have a chance, and must therefore be kept in appropriate places. It is not the tiger but the zoo keeper who deserves punishment.

Supposing that future financial markets with better institutional rules will be better than today's markets, one thing that nobody can promise is that there will be no more crashes, and nor will it be easier to make money. Safer markets will encourage investors to take on more risks, and some must have their hands burned to know the new bounds. We may ask the question: what is the purpose of a better market if it cannot eliminate crashes?

Future financial markets will reach a much higher base. Non-market economies are absent of crashes: North Korea may have famine and other calamities, but rarely financial market crashes. Setting up traffic lights and installing airbags in cars do not prevent accidents, and on the safest roads, fatal accidents will still happen. The Baghdad and London stock markets can both be marginally efficient, but efficiency is not a suitable measure of how markets perform. The main criterion will be that a better market will increasingly enable symbiosis between businesses and investors, which in turn will power economic growth.

Inherent instability can never be eliminated completely, as crash-free periods will embolden investors to commit ever larger sums to chase increasingly remote future opportunities; even the best market would still be marginally stable—an inherent feature of the capitalist economy. This does not mean that we must passively wait until a future global crisis eliminates irrational risk-takers. As with fining drivers for speeding, instead of waiting until bad drivers die due to accidents, in financial markets there must be preventive measures that punish careless investors earlier and more frequently before systemic risks cumulate to ignite a global crash.

6

Information Markets

This chapter extends market theory to markets of information and content. Although information exchanges are not monetary, the motivations of providers and seekers are important for information intermediaries in designing better platforms. Just as with previous markets there is the similar challenge of determining the quality of information and content. This chapter shows that information consumption models can be divided into three categories: searching, farming, and feeding. Some are more efficient and some more diversifying, and their advantages and disadvatages are discussed.

6.1 Signal vs. noise

Information can be a product per se, besides being a tool for evaluating other products, and in this section we deal with the quality issues on information.

Products and services can be misunderstood by consumers due to their cognitive fallibilities. Poor understanding results from natural noise—people can err. For example, evaluating a dish is easier than assessing the long-term effects of a food additive, and we may say that natural noise is stronger in the latter case. Our infocap will never be sufficient to match the bewildering array of information tasks around us, and we shall forever live with the natural noise. In general, natural noise is particularly acute for hard-to-understand information like that in finance.

Wilful distortions are called "intentional noise." In Chapter 1 we identified the defensive and cooperative regions across the infocap range, and discussed clever marketing gambits as like artificial complexities aiming to add confusion on top of natural noise. Intentional noise misleads people into perceiving what the sender wants them to perceive. Type "cheap Rolex" or "slim diet" into a search engine, and the most visible offers are of dubious origin. Information on commercial subjects is often plagued with intentional noise, more so than information on philosophy, astronomy, and so on. Natural noise makes finding objective information hard enough, but information providers may go to great lengths to distort the truth. Consumers are already naturally fallible; now we must reckon with the manufactured fallibility that is created deliberately.

Producers need to transmit messages of their products to consumers who are poor receivers, and markets are noisy transmission channels. Searching and evaluating are like decoding, whereas marketing is like encoding; moreover, encoders are specialists and decoders generalists. Vendors encode the original message of objective facts with some distortions (Section 1.3) to instigate consumers into buying.

Encoding or distortion is not necessarily always malicious. On a banal level, for example, a woman who is already naturally pretty may still resort to cosmetics to improve her looks. We do not always speak of distortions, but it becomes an accepted social norm. The same may be said of advertisements for genuinely good products, which is why businesses need to invest in both substance and appearance. Encoding sometimes takes on more professional aspects; expensive PR firms and spin doctors embellish a product or a politician, overdoing this might lead to abuses.

Search engines are far from being perfect in determining content quality, and webmasters might manipulate their sites to please them rather than real users. The industry of search engine optimization (SEO) aims to professionally help webmasters to improve their pages' appearance. A website's ranking has great economic consequences, as it can often make or break a business, depending on how high it ranks on search engines.[1] Some search engines use the practice of paid inclusion—unabashedly promoting websites without merit if they pay.

With a large indiscriminating audience, the days of intentional noise are not counted. Intentional noise does not have the same effect on everyone, but serves effectively as a filter to sort the informed from the uninformed and the experienced from the inexperienced.

We must often rely on experts and insiders for all sorts of evaluation, but experts are often the most likely to distort. They are a minority, and hence are easier to corrupt. Misunderstanding caused by experts is more intentional than natural, whereas misunderstaing by consumers is more natural than intentional (Section 2.3).

Now let us consider the third type: herding noise. Sometimes a chancy or intentional first move starts a herding momentum. For example, book agents buy their own books[2] in the hope to initiate a mass-buying trend. The herding noise rarely stands alone, and it often builds on top of the other two types; it is prevalent where natural noise is high. For example, cultural products such as movie scripts are hard to evaluate, and content brokers struggle to find the next megahit. Venture capitalists also have a hard time in identifying promising startups, and people often all pile on popular choices according to the current consensus.

The finance gurus' prediction of the ensuing year's stock indices is worse than what natural noise would imply, as their views tend to herd together. Financial markets are plagued with all three types of noise, of which herding noise is probably the most serious. Some can profit from initiating a herding momentum; shrewd traders might profess ignorance of fundamentals, but can take advantage of fellow investors who fall for the herding noise.

In a well-known paper, Barabasi and Albert[3] considered preferential attachment as the growth mechanism of networks. They noted that if a webpage already has many other pages pointing to it, it will attract still more pages linking to it. Preferential attachment is also known as the rich-get-richer scheme in many contexts, and we consider it a typical manifestation of the herding noise.

If initially a webpage by chance becomes slightly more popular, search engines might assign a higher weighting to it. Consequently, webmasters and users will have a greater chance of finding it and thereby linking their pages to it, which in turn will further increase its weighting. Webmasters often know what to link due to their offline contacts and expertise, but a fraction of their knowledge derives from using search engines. Information is acquired from both offline and online sources; the former tends to make the long tail longer, and the latter makes it shorter (Section 4.2). The rich-get-richer mechanism systematically biases against smaller, newer sources.

Google's PageRank algorithm and its variants accentuate the herding noise and leave behind a web with "popular" sites over-represented. Here the so-called "popularity" is due more to herding than to merit. Search engines might, however, pretend that their role is neutral and that they only meet users' needs. A significant part of the World Wide Web we see today has grown during the rise of Google, which has had a hand in growing a web that is systematically, though inadvertently, distorted with its algorithms. Most search algorithms today are based on rich-get-richer schemes, and over time they make the finding of long-tail items ever more difficult.

People behind intentional and herding noises can profit from them. For example, a marketing agent might pay celebrities in the hope of herding their fans to its business clients, just as Twitter vows stars with thousands of followers. Many information services' stated purpose is to help consumers to be better informed—and yet they intentionally distort information a little, and profit from doing so.

In fact, free information is never really free, as its peddlers encode, directly or indirectly, commercial messages for their paying clients. Google for example, does a good job in reigning over intentional noise by others, and then adds a little of its own in the form of sponsored advertisements. But win-win-win outcomes can still happen, as Google's help far outweighs the slight inconvenience of the advertisments, hence it can still enable overall positive sum games.

6.2 Decoding

Finding relevant information is one of the most challenging tasks in modern society. To fight the three types of noise, sheer computing power is not enough, and refined algorithms are necessary for finding the proverbial needle in an ever-larger haystack.

Natural noise is uncorrelated in general, and there is a simple way of finding the truth: the wisdom of crowds. There is a old anecdote that when many people were asked to guess the weight of an ox, the average of all the guesses was close to the true answer. The modern-day equivalent is to take average ratings on books, hotels, and products. The other two types of noise, however, might result in the ignorant majority making the wrong choice, and simple averages might no longer be effective.

AltaVista was the dominant search engine for the first decade of the Internet. Basically, it took the frequency of keywords as cues for quality and relevance. Webmasters soon discovered what AltaVista was looking for, and after they had inserted many dummy keywords in their sites, search quality quickly deteriorated. Google dethroned AltaVista by better combating intentional noise. The importance of a website is no longer determined by how often a keyword appears on a page, but by the PageRank algorithm based on the linkage among pages. To rank high you need other high-ranking pages pointing to you. Thus, millions of webmasters are effectively the unwitting foot soldiers for Google's billion-dollar business, in the sense that they choose links carefully—and Google's success rests on determining the significance of those links.

What search engines face is emblematic of the information challenge that ordinary people face in daily life. For example, on TripAdvisor, when we see a hotel with a good rating, experienced and diligent users will not just take the star at its face value. Ratings are easier to manipulate than detailed reviews. We can click further to check the people behind a five-star review to find their previous contributions for clues of reliability and relevance; in fact we do a mini-calculation, just as Google does on a global scale. We often initiate a mental PageRank algorithm on a limited scale; for example, when a friend who has previously made pertinent remarks suggests a new book, her suggestion might be taken more seriously than a suggestion by a stranger.

Ironically, intentional noise can be caused inadvertently by noise-fighting strategies. To evaluate and improve medical services, health authorities invented the system of report cards, with hospitals reporting success/failure ratios of bypass operations.[4] This had the opposite effect of improving patient care, as the hospitals under evaluation often turned away critically ill patients for whom a desperate operation might be the last resort, and preferred to operate on relatively healthy patients whose need was far from being compelling.

In China there was once a food scandal when a baby-milk powder was found to be doctored with a potentially deadly chemical. Dairy producers in China routinely dilute milk with water, but ingenious quality-control agencies designed a way of detecting the protein content in milk, rendering the relatively harmless water-adding gambit ineffective. However, the company Sanlu devised an even more clever idea of adding melamine (which can fake protein presence) to diluted milk so that its toxicity would not be evident until much later.

Intentional noise can be still more lethal than doctored food. It was reported that the Colombian army devised a kill quota for its soldiers fighting FARC (anti-government guerillas): if they reached the target they would be rewarded. But the soldiers found that whereas FARC was difficult to hunt down, unarmed farmers were easier substitutes to fill the quota.[5] These examples demonstrate that without a detailed understanding of human reflexivity, quality-control measures to combat noise can be counterproductive.

There are, however, ways to reduce herding noise. For example, by cutting reflexive feedback loops, online media might not allow voters/raters to see results before voting. However, this not-show-ratings measure is irrelevant for web growth, as we cannot ask webmasters to abstain from using search engines before choosing their links. Without radical innovations like that of Google's dethroning of AltaVista, we expect search quality to decline as accumulated distortions due to herding worsen.

Even search engines only aim to search existing information, they cannot ignore how the web grows; and Google especially has had a hand in building up a large systemic bias. In combating intentional noise, Google's algorithm inadvertently amplified the herding noise, so in solving the world information search problem it created a new problem that is more difficult to solve. The web receives reflexive impacts during its growth, and engineers are either part of the problem or of the solution, as they are an inalienable part of the information ecology.

Preferential attachment is an overlooked plague that amplifies the herding noise which in turn distorts the web; it damages the long tail and is resistant to treatment. The remedy will be costly, but one way out for future search engines is to consider another dimension: time. They may trace the information genealogy tree to get a deeper understanding than the static network now in use. Future search engines would determine, from growth history, how an original meme transmitted and proliferated, and this offers the best chance to undo the systemic biases caused by preferential attachment.

Another method of combating herding noise is to amplify minority opinion. Consider Amazon's book reviews. A book typically has several hundred pages, and it takes considerable effort to evaluate it. Reviews, on the other hand, are much more concise, and they may act as executive summaries. A hurried reader might quickly assesss a book by reading reviews, and some of these may reflect his own opinions. If he feels a review is helpful, he may vote for it. Informed reviewers, though in a minority, can have their opinion amplified, and their reviews, if well written, can sway many interested readers.

Andrzej Nowak, a Polish-American psychologist, mentioned a study of polling US presidential debates that might appear puzzling. Media companies are eager to poll the public after a debate to know who might emerge as the winner. Usually, however, they do not call people immediately after the debate, but allow them ample time to deliberate and poll them the next day to procure more reliable polling results. By deliberation within a small trusted community (within a family or among friends), the better informed might influence the less informed, and thus the minority opinion can be amplified. The more informed might quote background facts and offer insights so that the uninformed might form an opinion or abandon prejudices.

But do minority-amplification schemes contradict our goal of combating herding noise? Consider the following two scenarios: an ignorant majority swayed by a minority vs. another majority not yet fully understanding the implications but can be made so through deliberation. Social influence may not be a bad thing if the majority has retained its own judgmental power. For example, on important issues in a committee meeting, the chairman does not call for a vote immediately but allows deliberation, letting the informed minority expose the full implications. This hinges on the analytic power of the less informed majority. If they are totally ignorant, deliberation is of no use, but neither is it of any use if all are fully informed. Anywhere between deliberation can help the hidden infocap of the less informed majority.

It may appear puzzling: businesses, to an extent, spend much marketing effort (encoding) to distort commercial messages, and consumers must exert still more onerous effort in undoing (decoding), with limited results. What is the point? This may appear as pure waste. From the businesses' side they could not care less about global welfare, but what they do still makes commercial sense, as the encoding/decoding exercise can distinguish informed from uninformed consumers (Section 1.3).

6.3 From searching to farming and then to feeding

When we consider buying a new gadget or looking for a hotel, instead of querying search engines we often go to specialized review sites such as TripAdvisor for hotels, and Amazon for books and products. The choice of review sites is much narrower than the websites listed on search engines. As discussed in Chapter 2, reviews are written mostly by consumers, and platform matchmakers organize the content in a consistent format and impose a standard.

A new model of information consumption emerges that information can be created in designated places where users can easily find what they need without searching. This is called *information farming*, to be compared with information searching. Google and Wikipedia are the most well known examples of searching and farming. The metaphor of farming and searching for information is similar to that of farming and the search for food in agriculture. Farmers know how crops grow, and the harvest is just the last step (equivalent to what search engines do now).

Many signs show that the farming model is beginning to challenge the dominant searching model. When querying Google, we often find Wikipedia's answer at the top of the search results. But some people may have already changed their habits, and instead of searching the web with random results they have favorite places to go to. Wikipedia editors curate information with a consistent look and an air of authority. To a casual user this provides a quick overview, as its statements are backed up with authoritative references.

Efficiency gains of farming over searching are the ease to find content in a well-known place, consistence in quality, and stimulation of content creation. The first two are obvious, while the last one is the most important but is often neglected. When contributors to a given subject are connected they can measure their knowledge against the state of the art easier than when they are isolated.

When people contribute content to a well-known platform such as Wikipedia, both the pioneers and laggards can benefit. The laggards learn how far behind they are and should either do it better or do something else, while the pioneers might feel encouraged in knowing that many people appreciate their leading role and that they might do still better.

Farming does not involve simply rearranging pre-existent things; a matchmaker's farming role goes beyond that by laboring the ground purposefully for content to grow. Information farming needs the understanding of contributors' motives, and how the content is presented and appreciated can have reflexive impacts on contributors.

In Chapter 2 it was mentioned that we are frequently free-riders and occasionally contributors. On a typical item in Wikipedia only a very few are true experts, and the editing rules are not as democratic as might seem. For each item there is a discussion board where the regulars socialize, and they may form the so-called "quality mafia." Everyone can edit or deface an item, or so it may seem, but there are many invisible barriers where only a select few can influence the content substantially, and where the minority's opinions are thus amplified. As Wikipedia becomes more mature, for occasional editors the barriers are still higher.

Among core Wikipedia editors, many are passionate pseudo experts, as they acquire content mostly from other online sources. However, original knowledge comes more often from life spent offline than that online, and due respect must be given to those who have little time for Wikipedia politics. Some complain that lifelong professional pursuits are less valued than nightlong editing wars, and information peddlers of the secondhand or thirdhand often overrule firsthand creators.

The concepts behind farming can also help searching. Google's self-proclaimed mission is to organize the world's information—a motto which implies that Google takes a passive stance regarding information creation. Are you not happy with what you find? Sorry, but that is the best we can find on Earth.

Search engines can go upstream in the information ecology: the existent content is just a small fraction of what people can provide and ask. Just as we have extended the narrow supply/demand relation to include hidden potentials of consumers and businesses (Section 1.4), in the marketplace for information there is also much more than meets the eye.

Future search engines would adhere to a new paradigm: how to find information that has a direct impact on its creation. For example, they could enable dialogues between content providers and searchers. By watching such exchanges they would be able to determine better their users' hidden needs and providers' hidden capabilities, and would then indeed take an active part in content creation. For example, they might enable searchers to leave comments and feedback on webpages, and the dialog would then stimulate authors to improve their content. Information content is not about pre-existent objects waiting to be unearthed, but is socially created under the mutual reflexive impacts between creators and users.

This would be a dramatic departure from the current search paradigm, as previously non-existent content would more probably be created in desired places. The content-owners would be notified of visits to their pages, search engines would become moderators of the conversation, and the boundary between searching and farming would indeed be transcended.

Searching for information is initiated by users in need, whereas farming's emphasis is at the other end: the creators, and the two extremes should complement each other. Does this mean that in the future, information will derive from farming instead of from searching? If true, this would reduce spontaneity and even hinder creativity. Repeatedly used content is suitable for farming, innumerous items deep down on the long tail are still accessible via searching.

A more efficient information consumption model is called "feeding." We find information more easily in the farming model than in the searching model, but users must still fetch content. The feeding model is the easiest now that the content comes to us unsolicited, but hopefully they would be relevant. Ease is enabled by harder work by information purveyors, as in the discussion of the role of Personal Assistant.

Elements of the feeding model are already embedded in some web services. While we search on Google, it inconspicuously feeds us its sponsors' advertisements. In Chapter 2 we discussed extensively the recommendations of matching products to people. Information by searching and farming is rarely personalized, as feeding requires a high degree of personalization, while mass-feeding programs such as banner advertisements and television commercials are on the decline.

In Chapter 3 we discussed the pull model gradually replacing the push model, but the feeding model actually represents more pull than push. We may be as lazy as before, but our implicit and inferred wants are taken care of, as if each of us really pulls information; our PA's make us better pullers. Search engines simply rank items which they have indexed, but farming requires that matchmakers do more by helping grow content, and feeding requires the hardest work, as matchmakers and PAs need to follow users' information consumption history in order to identify their taste-mates.

In hunter–gatherer society, the moment that man felt hungry he would forage for food, but with uncertain results. Agriculture changed it all, and man learned to plot the ground so that crops might grow regularly. In the current stage of the Information Revolution we still live in what is equivalent to a hunter–gatherer society on the verge of transiting to farming. Man used to consider foodstuff pre-existent; he could not care less about how forest berries emerged, and good gatherers knew better than others how to find them. Search engineers are just like good berry gatherers, as they know how to find information in the jungle of the World Wide Web.

Jade Diamond, in his bestselling book,[6] emphasized the importance of agriculture, which ushered unprecedented economic growth, unleashed productivity, and lifted man from the gatherer–hunter society to an age of enlightenment. Information farming is in its infancy, and information feeding has only just started. Their role—which at the moment is very small compared to searching—is gradually being recognized and embraced, and their transformational impact on society can be spectacular.

The farming and feeding models fully recognize human resourcefulness on both sides: users and contributors. Instead of merely suffering from its impacts, they can turn reflexivity into a productive tool. A better design will bring a better harvest for both information farming and agriculture.

6.4 Informational diversity vs. efficiency

To emphasise that these information problems belong to the apple class rather than to the Apple class (Section 4.5), we now examine why we need to be wary of the unbridled pursuit of efficiency at the expense of diversity.

Search engines help us reach far and wide, and whatever innovation they divulge would promote information diversity—or so we might think. For example, Google Instant is touted as an efficiency-enhancing feature. When the first few words of a query are typed, popular items appear that might prompt a casual searcher to acquiesce to a suggestion. The engineers might save us a few seconds per search—a big improvement in efficiency—but Google Instant unwittingly harms the long tail.[7]

Engineers might retort that searchers have the freedom to ignore suggestions and instead type long queries. If people could exactly pinpoint their preferences, searches should indeed indulge in these only, but people cannot articulate exactly what they are searching for, and suggestions of popular items will narrow the choices. In a way, Google "hijacks" uncertain queries to popular items. It can create a vicious circle; as once the herding effect starts, popular pages will become further enhanced, with the fair chance of other sites being eliminated. The engineers' unbound zeal for efficiency and innovation can harm information diversity.

Google Recipe is a tool aiming to help people find recipes. Google engineers choose parameters that are easy to quantify—for example, cooking times, calories, and ingredients—to make searching for a recipe a more efficient process. The *New York Times* reported that original recipes are hard to quantify, and hence they tend to be ignored.[8]

Google Personalization may be a misnomer. It tracks your search history and anticipates that your future searches will be in the same niches as before, but the result is similar to the rich-get-richer scheme mentioned previously—that the longer you utilize past queries, the harder it is to acquire fresh items, as the weighting becomes more reinforced. What might have been a slight, temporary penchant for some interests might now trap your future searches. Stumbleupon.com does exactly the opposite: it aims to move out of your past taste trap.[9] Although Google achieves a higher precision (efficiency) to charter to your tastes, StumbleUpon enhances diversity.

Man knows more than he can tell (implicit knowledge), and his needs are still broader than what he knows. That is why we should relax the urge to be precise, and instead procure a healthy dose of "off-topic" serendipity. Moreover, we also need to go beyond our own implicit wants to be exposed to something completely new without justification—just for novelty and surprises. When browsing in a library or bookstore, we often long for something interesting; and likewise, browsing on the web cannot be replaced entirely by precision searching. Traditional newspapers allow more off-topic items to catch our attention than do digital media, which are designed to focus, apart from a few advertisements to distract our attention for lucrative purposes.

Information should be multifaceted and context-dependent, and it is rarely neutral. This is at odds with the neutral point of view (NPOV)—a fundamental principle of Wikipedia. Different people have different needs regarding a given item; a neutral item can be too bland to many. Wikipedia items are ostensibly neutral, and its designers hope that the best compromise will suit all, just like dishes served at Olive Garden restaurants (Section 4.4). However, there is seldom a single best answer to a question, and complex issues need to have multiple perspectives to produce a fuller view. Wikipedia, by design, aims at only one answer without questions, comments, and personal feedback. If questions were allowed on Wikipedia, user–contributor exchanges would make the otherwise bland information lively and rich. Quora allows more personal views, and its items are typically signed by contributors (often well-known experts). For users in a hurry, multifaceted information is less efficient, but it helps information diversity.

For instance, TripAdvisor's ratings (stars) are easy to understand but they can be readily manipulated, while travelers' reviews and photographs about a hotel will provide a more intimate picture and are harder to manipulate than are the stars. Reading original and lengthy reviews in toto is not an efficient proposition, but it might offer a more pertinent insight. Reviews such as those on Amazon and TripAdvisor offer the opportunity to find echoes in other people's experience and to draw personal conclusions.

Agriculture provides the bulk of our food, and wild crops supplement it with refreshing diversity. For the information analogy, on the efficiency hierarchy of browsing–searching–farming–feeding, the left end represents diversity and the right end efficiency. People with casual interests want simple and fast answers, while the more informed need diversity and depth that efficient systems usually lack.

For an information purveyor, the higher it is on the efficiency hierarchy the warier it should be of its impact on diversity. Browsing web pages is the least efficient way of accessing information, as the web is like a labyrinth in an unknown town. Blogs are more diversified than mainstream media, as they are more opinionated; readers' feedback enriches the original content, and diversity is naturally sufficient. Searching is more efficient than browsing, as search engines help us to focus. Farming makes the focus sharper, as we need only go to a few places, while feeding is the most efficient model, as information comes to us. As discussed previously, all the efficient models have their caveats, and if not handled with care, diversity would suffer.

Steam engines and search engines are quintessential examples of the Industrial Revolution and the Information Revolution. Both need engineers, but the mindsets should be different: mechanical precision vs. man's reflexivity and socioeconomic complexity. Just like social scientists, engineers also fall easily for the fixed-target fallacy.

In conclusion, we have showed that methods of searching information influence its creation, and that the ruthless pursuit of efficiency might be detrimental to information diversity. If search engineers would consider the bigger picture, including themselves in information ecology, they might be able to amend their negative impacts (such as herding noise) and promote positive impacts by taking a proactive role in content creation.

PART 3

MARKETS AND INSTITUTIONS

7

From Markets to the Economy

This chapter summarizes market theory and compares Natural Selection in biology with the new mechanism: informational selection. Besides many similarities, informational selection is generally weaker than Natural Selection, but it can be improved in various ways, and institutions are the most effective in helping individuals to deal with its challenges. It also has unique strength, in that agents can have the foresight to skip many steps in economic evolution. Informational selection is therefore the combined result of foresight and selection.

7.1 Informational selection

We have seen that selection plays an important role in markets. It is often compared with that in Nature, so to distinguish it from Darwinian Natural Selection we shall call it *informational selection*. What we have discussed regarding inadequate information capabilities in the markets can readily be extended further. Economic agents face multiple alternatives, and what appears to be an inevitable sequence of events is often just one of many possible scenarios which happen to be selected.

Informational selection is ubiquitous in our economy. Beyond the examples of the three types of market in the previous chapters, the book edited by Granovetter and Swedberg[1] includes many cases of informational selection between suppliers and purchase managers, between employees and executives, and among firms. In general, consumers choosing products, engineers pursuing innovations, company boards selecting executives, managers evaluating employees' performance, investors sifting through data for better deals, school principals recruiting teachers, parents choosing schools, and every now and then the electorate electing politicians, can all be regarded as instances of informational selection.

Informational selection (infosel) and infocap are closely related but are not the same. For example, although weapons play a key role in battles, a better organized army might still defeat a better equipped army. In a famous essay,[2] the American anthropologist the late Clifford Geertz described in vivid detail how a bazaar economy operates, where sellers and buyers deploy cunning schemes about products' information. Although his description concerned a Moroccan peasant market, similar information challenges persist in modern markets where products and services are more complex.

In the simplest case, infosel and infocap happen to be the same when a single buyer deals with a seller one on one. As soon as there are two buyers and they talk to each other, infosel will be different—depending on whether or not they exchange tips. Improving infocap leads to stronger infosel, but the most significant boost comes from designing better institutions such as the matchmakers discussed in Chapter 2, and especially Personal Assistant in Chapter 3.

While Natural Selection works tirelessly day and night, informational selection is sometimes on and often off. Natural Selection can detect the slightest advantageous variations, but infosel might overlook many superior choices. As many innovators know, it is not sufficient to have the best idea; they must reckon with human selectors who are fallible and biased. Because of weak informational selection, rewards do not faithfully represent merits, and appearance often deviates resources from substance as our economy becomes more complex.

Informational selection faces an additional hurdle: manufactured fallibility. Humans are on both sides of selection; for example, the selected side (businesses) might manipulate the selecting side (consumers) to degrade the latter's infocap. It is part of our daily experience that the good might sometimes be ignored and the mediocre might succeed. Human agents often game informational selection successfully, while animals cannot fool Nature. The invisible hand à la Adam Smith works better in biology than in the economy.

Informational selection is not limited to economic activity, as it operates wherever humans deal with each other. Different disciplines face different selection pressures. Chicago University psychologist Mihaly Csíkszentmihályi's book[3] *Creativity* discusses quality criteria of social and natural sciences. He notes that it is easier to judge a theory in hard sciences than in his own discipline of psychology, where it might take a lifetime to appreciate what is truly good. Alain Sokal, a mathematical physicist, stirred up a scandal when he published an article in *Social Text*—a social science journal—and then revealed that he had written a hoax by chaining many fancy words in a frivolous way. With this prank, he intended to expose lax criteria for quality in some social sciences. Our conclusion, however, is different from that of Sokal—not that some sciences are superior or inferior, but social disciplines are more complex, and quality is not as easy to ascertain. Informational selection simply faces a much harder task for social sciences. Indeed, the strength of informational selection varies greatly across different professions, and it is far easier to evaluate a chef's dish than a philosopher's new ideas.

Among economic activities, informational selection faced by traders on financial markets is probably the harshest, and it resembles merciless Natural Selection; the trader's motto "you are only as good as your last trade" tells it all. Plausible superior theories have little glamour on the trading floors. And yet infosel in finance is still far from being perfect, and it took several years to understand the Internet bubble and the subprime crisis. On the other hand, high theories in economics and their proponents, can enjoy academic esteem for a lifetime, and it matters little whether or not they are verified.[4]

It is not always a bad thing that informational selection is weaker. We do not perish when we are not at our best, and we often have second chances that are unheard of in Natural Selection. Informational selection is particularly weak; for example, when choosing teachers, doctors, or philosophers, as compared to choosing cooks or traders, performance measurements are necessarily more vague. Failures are milder in the economy than those in Nature. Bankrupted companies can recycle themselves in order to try other plans whereas creatures in Nature fail by death only. Our economy leaves a catacomb of failed ideas, projects, and companies, but it is not littered with corpses.

7.2 Selection and amplification

Selection pressure acts as the driving force for economic growth.[5] Tiny but ubiquitous actions such as evaluating a product and posting a review are examples of innumerable ways of contributing toward better selection of consumer products. Tibor Scitovsky noted that it is collective selection pressure that keeps vendors relatively honest. The selfish driving force can also bring win–win outcomes. For example, informed shoppers shunning bad products can also benefit fellow consumers.

The driving force may not always be economically motivated or even have a motive, and yet it can have huge economic consequences.

Better informational selection not only enables more consumers' wants and businesses' offers to be matched, but also acts as a filter to allow better products to expand, while the cumulated effects lead to overall economic growth. The selecting side is also impacted; for example, skilled investors will become wealthier and have greater influence on the financial markets.

Information in the economy or in society is inhomogeneous, as each of us knows something better than others. It is instructive to consider a minimal set of items about which everyone in a community knows, and a maximal set of which any item is known to at least one member. Thus, the minimal set is the common denominator, and the maximal set delineates the collective cognitive boundary. The community in question can be all the employees of a company, where the minimal set may represent some basic facts of the company. If the whole economy is under consideration, then the minimal set is very small.

For items inside the minimal set, information can be regarded as nearly perfect. It is probably no surprise that economics textbooks often quote apples or oranges instead of more complex products such as mobile phones or vacation packages that easily go beyond the minimal set, hoping that on such basic items nobody would have information deficiency.

Mainstream economics may be valid for the minimal set, but the vast gray zone between, where the real-life economy operates, is the most interesting. The American novelist William Gibson has famously stated: "The future is already here—it is just unevenly distributed."[6] If we make the minimal set larger, we inevitably also extend the maximal set, as any system-wide improvement inadvertently expand, cognitive boundaries.

We shall forever live within the gray zone of knowledge, which gradually extends into the unknown. The driving force of informational selection operates in this zone, where wealth can be created. Mainstream economics barely acknowledges any driving force, as it is concerned with a final equilibrium state and has scant interest in the process of getting there. In sharp contrast, we focus on endless processes generated by the driving force.

Institutions also play an important role in boosting the driving force. Douglas North once pointed out that institutions can reduce uncertainties.[7] Well-designed rules, customs, conventions, and regulations can serve as substitutes for the driving force when scrutiny is repeatedly solicited. Consumers rely on regulating and certifying institutions to keep vendors relatively honest, and a few experts at FDA can spare much effort of millions of patients.

In a *New Yorker* article, the surgeon and author Atul Gawande tells how checklists for intensive care units and among ace pilots in the Second World War can save thousands of lives.[8] Surgeons and pilots are able experts, but their checklists still show superior results with less effort. Checklists may relieve them of the routine burden, and they can focus on the unexpected. Traders likewise need back offices to keep track of their trading positions, allowing them to concentrate on identifying trading opportunities.

The driving force maintaining the status quo can often be replaced by institutional rules. It is like digging a tunnel; immediately behind the advancing drill, workers set up supporting structures. Digging and supporting are indispensable for building tunnels; the former probes new territories, while the latter consolidates the advances. Quality of all sorts in our modern economy seem acceptable even without our apparent diligence, and many institutions keep things in order in lieu of the driving force.

The speed with which the driving force makes impact can vary greatly. In Section 1.2 we mentioned that under selection pressure a firm can dig deeper into its resourcefulness. In the real-world economy, there are many imbalances for which corrections will last a long time or forever. For example, wage differences between genders have historical, cultural origins.[9] When oil prices increase rapidly, one might expect that market forces will accelerate the supply, but a typical oil production cycle can be a multi-decade undertaking. Equilibrium economics is rooted in a misleading analogy: pendulum movement. As soon as it is knocked off balance, restoring forces will push it back, hence there is little interest in the process.

7.3 Who creates the wealth?

In the economy we do not simply rearrange and allocate resources; someone must create them. Moreover, there must be constant wealth creation to offset massive waste, corruption, and losses due to natural and man-made calamities. In Chapter 1 we identified magic pie as the basic wealth-creation mechanism, and here we shall identify the agents. It appears that some altruists have contributed more than they are paid for. Who are they, and why?

Individuals make a living by providing their skills for the economy, but their rewards are not necessarily in proportion, infosel being weak. Although some are truly altruistic, most us may still be self-centered. Does this imply that everyone would give less and take more? We shall see that the opposite is often true—hence the apparent puzzle.

It is easy to understand why there are many opportunists, but it seems paradoxical that there are far more altruists. Here we argue that people often cannot appropriate all the value they create; they are generous altruists, despite themselves.

A shoemaker can charge for a pair of shoes whatever a buyer is willing to pay, and here mainstream economics enjoys the benefit of doubt that the value the shoemaker created is fully repaid, that his reward may be exactly equal to his input. However, most of our skills are not marketable, as they are evaluated by fallible humans. For example, people writing a book or inventing a new algorithm cannot easily make a direct exchange with those who may benefit from it; a software engineer cannot extract reward from codes as simply as a shoemaker can from shoes. Computer codes, although valuable, are hard to price. As our economy becomes ever more elaborate and complex, there will be more jobs for coders than for shoemakers .

People either over-contribute or under-contribute, and contribution-equal-to-reward is more an exception than the rule. Professional skills are often assessed within an organization by non-monetary measures. If your contribution is more than your wage, it is because that there is also an additional value creation in coordination by your firm. An yet the firm does not just hire as many over-contributing employees as possible, for it also faces a similar dilemma: the surplus generated by their employees may not be fully appropriated, for it must stay competitive.

In a way, resourcefulness creates the value inside a firm and competition delivers the value to the outside economy.[10] Competition forces a firm to disgorge a part of the surplus from its employees; a monopoly position certainly helps, but pricing power also depends on how easy it is to charge end consumers. Innovative firms leave more on the table for consumers than do efficiency-chasing firms. On the weaker side of information asymmetry, consumers often suffer fraudulent tricks, but also enjoy unexpected bounties that businesses disgorge despite themselves.

The differential between the value created and the reward received can be beneficial to the economy in countable and non-countable ways. If something cannot be priced it does not make it less valuable, though it may or may not appear in the GDP. What otherwise must be paid for (products and services), now beneficiaries procure it cheap or free, and the savings can be spent elsewhere or passed on. For example, many believe that Craigslist operates far below its full monetization potential (Section 2.4), as its free or cheap listings enable users (individuals and firms) to make or save money. If Craigslist generates $1 billion in revenues each year, its much larger invisible value might register in the GDP with thousands of its clients, and still more may never be counted in any statistics.

Value spillover is hard to measure. For example, Google has an estimated ratio of two: each dollar of its earnings generates $2 elsewhere in the US economy.[11] Swiss beekeepers receive 60 million Swiss Francs per year for honey, and a study has shown that their incidental value spillovers into agriculture is much larger (600 million Swiss Francs)—hence we can speak of a spillover ratio of 10. But even though the beekeepers are only interested in honey, their role beyond the initial scope cannot be ignored. Positive and negative spillover ratios represent underlying positive and negative sum games, with the zero-sum game being the special case.

Partial appropriation makes many people over-contribute,[12] and many opportunists also get away with their ill-gotten profits. The overall result is a compromise between "wastes" versus generous "donations." The scandals and embezzlements exposed in the media are just a tiny tip of the darker side of the economy, which is apparently offset by our massive over-contribution that rarely receives due recognition. Both inadvertent donations and intentional cheats are the consequences of limited infosel. Still, a small fraction of the residual (positive) value survives and registers as economic growth. If infosel works well, our baser opportunistic deviants are discouraged, and positive deviants are promoted.

Weaker selection strength seems to produce a larger ratio (contribution vs. reward). We may ask: does a still weaker strength really imply more free giving? The answer is "no" (otherwise a clever idea, since weaker infosel is so much easier to achieve—sabotage!). In the extreme case of the selection strength being null, a person's reward has no correlation at all with his or her contribution, as in failed utopian economies.

Informational selection plays a trick on us: it is strong enough to motivate us to contribute, yet not sufficiently strong to allow full appropriation.

It is difficult to ascertain whether people ultimately become free-riders or contributors. The former may not necessarily want to be parasites, but they may become stuck in a free-riding position because of institutional settings that impede them to find more productive conduits. The latter may not always be altruistic, but they simply cannot appropriate what they create. In the case of a scientist or a philosopher, for example, even when praised for being successful in their respective areas their economic value to society can never be calculated precisely. But if it is over a long-time span and averaged over many sectors, our collective contribution to the economy and society appears to be a positive sum game. Our infosel strength, though limited, still often preferentially picks up the good and shuns the bad—and, of course, it has great scope for improvement.

The mainstream doctrine of reward-equaling-contribution is not only false but can also be harmful. Those taking more than giving by profiting from legal loopholes can whitewash unfair gains, while those over-contributing are not recognized. Denying the existence of under-contribution and over-contribution is to condone the former and to discourage the latter. In a Robinson economy, there are only natural elements to reckon with, and Natural Selection reigns—full appropriation. In our modern economy, appropriation is probably either over or under.

Defenders of the equilibrium hypothesis sometimes claim that apparent under-rewarding of individuals can be explained by non-monetary benefits such as pleasure and peer esteem. Consider scientists winning a Nobel Prize—such as in 2009 for the invention of optical fiber—for which the laureates' contribution is probably far greater than the prize money. There is obviously prestige and personal fulfilment conferred by the award, but can we really affirm that the non-monetary rewards equal the differential? Such a claim would be based on faith, and would forestall detailed inquiries into the real causes of value and motives of the individuals who create it.

7.4 Flexibility vs. commitment

Weak informational selection makes finding the best a challenge, but a way to boost selection results by adhering to a non-optimal choice, as the selected side may go for its hidden potential under the right circumstances. Insiders in financial markets can also have a sub-optimal choice in hand, but they can gain by exerting more effort to change the business that they own. The additional gain from tapping into their hidden potential may compensate for the loss due to poor selection.

Flexibility implies choosing or eliminating, whereas commitment means adhering to a choice to unearth its hidden potential. Since informational selection is limited, we cannot rely on finding the best. Prior to selection, agents choose among alternatives (flexibility); post selection, they can commit to the choice even though it may not be the best.

In Section 3.4 we discussed prior- and post-production stages, simply denoted as B2 and B1. We see that the B2 stage corresponds to commitment and the B1 stage flexibility. In consumer markets, "selection" traditionally means either picking it up or dropping it. But we see that the selection concept in markets is extended, and both commitment and selection are important. If only hard selection is present, businesses would produce blindly to an extent, and only once in a while would a good product happen to succeed by luck. Therefore, the commitment part (B2) helps considerably in bringing out the hidden best of businesses.

The modern economy has a vast transaction zone.[13] From raw material production to end consumers there are many intertwined layers and nodes, and many intermediate businesses and suppliers often deal with each other. This structure can be called a *value web* (or value chain). For convenience, we refer to the raw material end as upstream and the consumer end as downstream of the value web.

Let us take a closer look at the relationship among nodes within the value web. A firm does not choose a new supplier each time it buys supplies. In principle, markets offer everything it needs, but its limited infocap obliges it to concentrate on a handful of suppliers. Instead of businesses occasionally dealing with each other in open markets, the economy has a recognizable network structure. The structure of the value web is neither rigid nor fluid; it is made of what Stanford sociologist Mark Granovetter and his colleagues call "semi-embedded relationships."[14]

Geertz has also observed that the bazaar traders relied on "clientalization" (durable relationships) to compensate for their limited infocap. Business people constantly struggle to find the right balance between flexibility and commitment to adapt to their limited and changing infocap. A stable and committed supply relationship brings many advantages, as it allows partners to engage in frequent transactions when contractual details cannot be adequately stipulated in advance.

Brian Uzzi, a business school professor,[15] followed the garment industry in New York for a year and revealed its fascinating, complex, and mutually supportive relationships. His conclusion was that with semi-embedded relationships the mutual pie is larger than that derived from market transactions. He found that commitment cannot be too strong, lest partners forego any possibility of selection. Firms should always hedge their bets by keeping alternatives open and to keep partners on their toes. The relationship should be sufficiently stable that occasional failures can be tolerated, and suppliers may even go out of their way to help partners in difficulty, instead of dumping them at the first disappointment.

There are two reasons why an agent does not want a completely fixed relationship with partners. The first is that he is not sure he has the right choice, since he does not have the time and infocap resources to check all the others. The second is that a fixed relationship leads to complacency, as the suppliers might not have sufficient incentives if they know that they will not be dropped.

Ronald Coase first recognized that the existence of firms is a way to overcome "transaction costs" in production organizations. Using our terminology, we shall replace "costs" by "information difficulties" or "limited selection strength." Relationships within a firm are more committed than semi-embedded relationships among firms. But even inside a firm the constituent parts can fine-tune the flexibility vs. commitment balance, and its size can shrink or grow by selling some divisions or acquiring others.

There can also be market elements inside a firm (incentives, bonuses for employees, and so on), just as among firms there are cozy embedded relationships. In recent history we have seen extreme examples of flexibility vs. commitment. Rigid structures have no selection and only commitment, like that of many failed planned economies in the last century. Pure market relationships exist only in textbooks.

The flexibility–commitment dichotomy can be portrayed as an information problem. Were infocap unlimited, no commitment would be needed either within a firm or among firms. The balance between flexibility and commitment is constantly shifting due to the influence of many factors; for example, policies and advances in information technology. Informational selection works best when it tackles adequate tasks, and mismatches between infocap and tasks will reduce overall selection results.

In Natural Selection, besides mutations there is also genetic recombination, and variants can develop from recombination and crossovers. What is intended for one purpose may not succeed but can have unintended outcomes for other purposes, and a failed project might later recycle itself for unexpected uses.[16] The flexibility–commitment dichotomy takes a new turn: committed activities can be for different reasons and then be selected for another function. In this case, commitment is still essential; only the devoted can delve sufficiently deep—like scientists pursuing a lead—and then the firm and the outside economy may select their inventions for totally different purposes. For example, chip and computer inventors never dreamed of current Internet trends.

Beyond the economy, informational selection can be just as important: across social, cultural, and political activities, relationships are often semi-embedded. We normally allow considerable margins of error and a rather long duration for a politician to produce results. For example, four years is a typical presidential or parliamentary term in a democracy, whereas forty years is typical of a monarchy or dictatorship.

7.5 Marginally efficient economy

We shall show that the economy cannot be made perfectly efficient no matter how hard we try—abd there are wonderful reasons for this. The following analysis is a continuation of the theme about new pies (Section 1.5), but now from the perspective of market efficiency.

In the economy there are always both margin-reducing and margin-creating or stabilizing vs. destabilizing forces. The more effective competitors chase innovators; the shorter is the latter's domination and there is more pressure to extend to farther horizons, and the new frontiers bring unexpected opportunities. New opportunities need time to be understood and exploited, and perfect efficiency is never the goal, nor is it attainable.

Inefficiencies in an economy are connected, and hence we cannot treat each separately. Just as we cannot drain dry San Francisco Bay, being connected to the ocean, any sector in the economy cannot alone be made perfectly efficient. Legend has it that Henry Ford (the founder of Ford) commissioned a study of junkyards to discover how car parts endured during its lifetime. All parts were found to be broken at one time or another, except for the kingpin (kingbolt). If it were an isolated task the engineers should be praised, but Ford considered that its over-specification contradicted the larger scope: the kingpin was too good with respect to the rest of the car. The zealots seeking isolated perfection were reprimanded.

Perfect efficiency acts as the benchmark in mainstream economics. Under such a doctrine, each transaction or each sector is dealt with independently. If deviations gravitate toward the optimum, why bother with the relationships among them? In the real-world economy, exploitable opportunities flow from sector to sector, and efficiency reduction or enhancement in one sector will drive out or attract agents to or from elsewhere, making the study of connected inefficiencies paramount.

This poses new questions that were non-issues for mainstream economics. Our economy has many competitors but also many sectors. Consumers' enhanced infocap will enable exploitation of previously non-viable opportunities (new pies), and it will open new frontiers—a process without end.

Physical efficiency has limits. In the case of the internal combustion engine, for example, each year some efficiency gains are achieved, but further improvements will necessarily slow down and will never surpass a theoretical limit. Economic inefficiency exploitation, on the other hand, knows no bounds, for we never have a fixed set of consumers' wants and businesses' offers. In the history of economic development, each time we seemed to hit an insurmountable wall the economy took an unexpected turn.

It is instructive to consider the following mining analogy. In the early days of mining, ores needed to be of high grade to merit exploitation. As mining technologies improved it became possible to extract useful elements from poor ores, but it is not economically viable to extract the last gram of gold from a mine. Because of diminishing returns it is not profitable to continue with a mine until it is exhausted, as improved technology can exploit mines elsewhere.

The new pie concept goes further than the mining analogy. Minerals on Earth, and any other physical resources for that matter, do not grow, but our new pie possibilities are truly unlimited beyond imagination. The zero-margin limit which serves as the equilibrium benchmark is not only dismal but also unattainable. Strictly speaking, the leftover margins labeled as "inefficiencies" will forever play a pivotal role in the economy; there is something wonderful about these inefficiencies that some create and on which others feed. In pursuing exploitable inefficiencies an inadvertent consequence is to speed up exploration of new horizons. We may never be able to separate exploitation from exploration, and we shall always deal with a marginally inefficient economy.[17]

The fixed-target fallacy often induces people to draw erroneous conclusions, either optimistically or pessimistically, with the same root assumption—that the future economy will deal with the same set of problems as we have today. Many thinkers fall for the nirvana-tomorrow trap. Alexander Pope[18] confidently predicted that in the future all would be clarified and that total clarity would prevail, but he neglected to consider that increasing knowledge and progresses will beget new opportunities and concomitant uncertainties. The same fallacy can also lead to unwarranted pessimism[19] that one day productivity growth will come to a stop or petrol will be exhausted.

7.6 Higher economy

Economic transactions are often enabled by middlemen—matchmakers, certification authorities, violation surveillance, copyrights, courts, and so on. By "higher economy" we mean all that encompasses the economic institutions which enable the transactions.

In our modern economy, transactions often take place without any obvious outside intervention. For example, the customers of a bank may be blissfully unaware that their dealings are safeguarded by many players behind the scenes, such as regulators, central banks, and surveillance agencies.

In real life, institutional structures can be more complex than the threesome relationships (buyers, sellers, and matchmakers) that we have used as examples (Section 2.1), but the threesome archetype captures the essential features that differentiate it from the twosome relationship that is typically portrayed in mainstream economics.

Institutions can contribute to economic growth by enabling and constraining the agents—promoting their upside (positive sum games) and plugging their downside (negative sum games). But these enabling and constraining roles are not contradictory, as they might, for example, impede profiteering in order to promote better welfare for the majority of people. The higher economy also includes enforcement, which can also bring value.

For example, we do not allow soccer players to sort out their rivalries among themselves, and the referee's role is crucial to the conduct of a good match. We hate traffic lights when they stop us, soccer players and coaches curse referees when fouls are called against them, bankers are discontent about SEC's long nose . . . and yet these higher roles interfering with businesses should not be abolished.

If laissez faire were to be dominant in severely information-asymmetric areas such as healthcare,[20] people would probably die because of unlicensed doctors or untested medicines before it were realized that it was time to change. Entrepreneurs are resilient and resourceful, but what they cannot achieve is higher-level improvement of designing rules and institutions safeguarding the business environment.

Fallibility is an issue on all levels, but on the higher level its impact can be more severe. Even if agents were infinitely capable they would still not be able to offset the lack of higher-level institutional oversight. One oft-heard complaint is that governments know no better than practitioners—so why let outsiders interfere?

Referees cannot play as well as can soccer stars, and traffic police are no match in facing F1 pilots; similarly, government employees in regulation institutions might be less competent than those in private sectors, yet the higher role is important not because of regulators' superior skills, but because of the special position that can facilitate better outcomes than can the agents alone.[21]

Of course, the higher role can do harm as well as good, but the imperfect job calls for improvement, not for its elimination based on market fundamentalist ideology. Higher-level intervention is often misunderstood and abused. Market fundamentalists deny it or write it off, while totalitarian regimes go to the other extreme by commanding the economy top down.

The bazaaris described by Geertz may be cannier than the shoppers in developed economies, but due to market institutions the latter may fare better. For the bazaaris, non-standard quality specifications make the hurdles high for informational selection, while in developed economies the problems become more simple due to regulations. It is the institutional settings that make us collectively more capable of dealing with information deficiency. With old tasks being relieved, in developed economies consumers may be less market-savvy, but they can devote attention elsewhere.

Informational selection operates on different levels. Examples of lower-level selection can be consumers choosing products or patients finding doctors. Examples on a higher level can be designing an eBay, a Craigslist, or a stock exchange. Informational selection on a higher level no longer handles transaction-per-transaction, day-to-day details, but uses foresight to enhance systemic quality-selection capabilities.

For example, car designers and drivers aim for safe driving, but they operate on different levels. Designing a car is to exercise foresight to reduce the likelihood of accidents, whereas driving a car is about safety at every moment. Road planning and city planning can be regarded as on a still higher level.

The same analogy applies in another context. In a developed economy (such as Switzerland), ordinary citizens are not necessarily more clever than those in a badly run economy (such as Somalia or North Korea), as the huge difference in productivity is due to their respective institutions. Agents on a lower level everywhere strive to take care of themselves, and scholars need not exhort them to be more rational. They already do what they can—sometimes under severe constraints. Policy-makers and scholars should instead concentrate on a higher level like that of car designers and road planners. Informational selection on a higher level should have larger scope and longer horizons than the agents themselves.

How do we compare informational selection on the higher level to that on the lower level? Birds and aircraft have strong and weak points: though aircraft cannot match with a bird's agility and efficiency, but they have unsurpassed range and power.[22] McDonald's is a well-known institution in the fast-food industry (its hygiene, for example, is evident in developing countries), but we cannot simply proclaim that its food is healthier than that on Geertz's bazaar. A higher economy creates new problems while solving better problems of a lower economy.

Institutions bring in more complex factors, and hence any attempt at optimization will be futile. Entrepreneurs are less concerned with perfection and more with seizing opportunities in the new space earlier than the competition. While matchmakers help other economic agents improve efficiency (buyers and sellers find better matches, for example), they leave themselves open to new inefficiencies (for example, matchmakers can only appropriate a small fraction of the value).

The transition from Adam Smith's twosome economy to the higher (threesome or more) economy has been a long process. The economy evolving to higher forms can be summarized: in the primitive economy, do it alone; in the economy *à la* Adam Smith, do it in two; and in the higher economy, do it in three or more.

We may admire the jungle laws for beasts in the wild but condemn them in the market economy—so why the double standard? This is another difference between Natural Selection and informational selection. When beautiful creatures such gazelles are devoured by lions, few people feel obliged to intervene. After all, natural selection over millions of years has taken care of life on Earth, and the result of its brutal forces is today's wondrous life.

For economic affairs, people call for intervention if poor people starve, banks fail, or nations are hit by calamities. Traffic planners do not let drivers spontaneously emerge into orderly flow and wait until bad drivers die out through accidents, starving people's bad fortunes may not be their own incompetence and they deserve second chances, and regulators do not let unlicensed doctors practice and allow patients to die.

Our modern economy is far more sophisticated and efficient than that of Adams Smith's time, but it is also far more fragile. Simple win–win outcomes are less likely to spontaneously emerge today than in the good old days when the invisible hand seemed to be competent. For the modern economy, most positive sum games are created and maintained by elaborate institutional designs, which are vulnerable to human fallibilities.

8

Man and the Economy

This chapter looks deeper into human needs and capabilities. We find that both are inexhaustible, while more are revealed and still more to emerge. Institutional settings are important in selecting and fostering desirable traits in man, and finally it is by expanding man's capabilities and needs that the economy grows. This chapter draws largely on inspiration from psychology, anthropology, and other humanities sciences, combined with our new market theory, to show that the theory finally finds its anchor in man.

8.1 Primary and secondary propensities

In this chapter we concentrate on the main actor—man—and take a closer look at man's motives and capabilities. Mainstream economics treats man's wants and skills as if they were given. Our theory instead considers that what meets the eyes of a statistician is just a tiny fraction of an inexhaustible pool. We show that when people become affluent their needs and secondary interests grow rapidly in number, and our economy will cater to more and more of these.

Our self-interest is supposedly universal, and all other idiosyncratic motives are hard to quantify. It is understandable that the founders of classical economics postulated man's self-interest as being as reliable in economics as is gravity in physics.[1]

We posit that while self-interest may be the primary motive, to understand the economy we must pay attention to many secondary motives. Although any single one of them can hardly rival the primary motive, collectively the secondary motives can be more important than the primary motive, and they are the true driving force of our economy.

Some economists criticize the self-interest/gravity analogy and consider a more complex man,[2] alternative theories such as behavioral economics recognize man's many failings that prevent him from being perfectly rational. While previous work exposes man's imperfections, our theory instead focuses on the productive side of man's secondary motives. The aim of this chapter is to show that secondary motives play an important economic role.

Mainstream economics considers self-interest as the only driving force, hence the notion of *homo economicus*.[3] Admittedly, man may be selfish; facing the clear choice of 10$ or 100$, even a poet would probably choose the larger amount. And yet even the most rational people still harbor secondary motives or do things without explicit motives.

If daily physical movements were due to gravity alone—vertical falls—Earth would indeed be a dismal place. Many sideways and gravity-defying movements—such as running, jumping, and dancing—make life interesting. Likewise, writing off secondary motives is the chief reason why mainstream economics was made dismal. Our theory focuses instead on idiosyncratic secondary motives, and we shall show that a fraction of them can be selected to extend our wants and skills, which in turn power the economic growth.

Recent research[4] supports the notion that most individuals are not exclusively self-regarding in their social interactions and that they are "conditional cooperators." Man often takes a first goodwill step of cooperation despite the risk of its not being reciprocated.[5] It is the gravity-defying motives that make cooperation possible. For example, Nobel laureate Elinor Ostrom's work[6] has shown that people are more likely to initiate goodwill cooperation if they know each, other than otherwise.

Widespread Internet applications challenge established views of mainstream economics. For example, millions of people contribute to Wikipedia and a plethora of similar websites—common knowledge repositories—without rewards, and public goods of knowledge emerge as if out of thin air. This leads to the debate concerning whether secondary motives can upstage the primary motive (amateurs vs. paid professionals).

For example, Nicolas Carr, former editor of *Harvard Business Review*, and Yokai Benkler, author of *Wealth of Net*, made a bet known as the Carr–Benkler wager[7] on what would turn out to be the dominant knowledge-generating model in the future. Carr derides user-generated content as an unsustainable fad, and Benkler insists that secondary motives have far more potential than has been evident so far. We need both professional firemen and a large pool of volunteers for firefighting, and it is probable that a mixed model would also suit content generation.

The primary and secondary motives are both important, and they complement each other.

The open-source software movement has an important role in IT with which businesses big or small must now reckon.[8] Thousands of coders with all kinds of secondary motives collectively produce high-quality complex software, and they often rival paid professionals. A famous example is Microsoft vs. Linux for computer operation systems, as described in Raymond's book *The Cathedral and the Bazaar*. Microsoft hires engineers, while Linux relies on an elaborate peer-to-peer network to achieve quality codes.

In the open-source model, no-one is obliged to work, and people do it for 'fun.' The 'fun' factor stems from secondary motives that include pride, peer appreciation, love of beautiful thingscreated, and bragging rights.

Not every profession is as joyful, and some jobs are known to have less fun. Janitors unblocking toilets and dentists drilling holes probably have less fun than programmers coding, and the moment the payment stops, the work may also stop. We do not open-source boring jobs to volunteers, but for many professions some fun can be had in work.

Tibor Scitovsky, in his book *Joyless Economy*, shows that man always yearns for something stimulating, and that absolute rest is undesirable and even frustrating.[9] Folklore illustrates this point: "A hen lays an egg, a poet writes a poem, and both feel better after having done it."[10] The American psychologist Abraham Maslow, author of the famous book "*peak experience*" raised similar points in his self-actualization theory. These are in sharp contrast to the Taylorism of a century ago, which urged workers to "check their brains at the gate" and to mechanically execute precise routines.

Shrewd entrepreneurs no longer focus on the primary motive only; instead they exploit idiosyncratic secondary motives for creative and productive ends. Recent views of man find that tapping into secondary motives will not only boost productivity but also foster human fulfilment, and the two goals are no longer seen as opposed to each other.

Man tends to be less calculating and has bigger pictures than do businesses. Clay Shirky, in his influential article "Fame vs. fortune: Micropayments and free content,"[11] shows that whereas publishers are interested in revenues only, authors have many other motives that may conflict with profit imperatives. Firms are short-lived vehicles, and they do not need to worry about the fate of our planet hundreds of years from now. Man may prefer not to grab until the last cent at the expense of the environment.

For the affluent, wealth is no longer about survival but about status, legacy, and power. Self-interest does not fade with a higher net worth, and the opposite is often true. The wealthy are more often involved in lawsuits and economic crimes than are ordinary salaried people; a man is more likely to cheat to acquire a yacht than another is for bread.[12]

Different parts of the world are in different stages of affluence, and the contrast is easy to note. In a Chinese village renowned for its hand-woven textiles and wax coloring that people have used throughout the centuries, many visitors now swarm to the place and pay high prices for such hand-made, defect-ridden textiles. Villagers gleefully cash in and then buy synthetic but perfectly made fabrics for their own use.

The Great Wall was built by Chinese emperors for a sole purpose: defense against invaders. It was in disrepair until only a few decades ago, and the local populace prized it as raw building material. Now, however, it is considered to be one of the world's greatest heritage sites, for multiple reasons—historic, national, cultural, and the desire to preserve an edifice which can invoke a sense of awe.

Man once considered other creatures only as food, but now, whether they are tasty or nasty, lovely or ugly, they are looked upon with tolerance and even admiration, especially in the developed world. The reasons are again multiple, but they are harder to articulate than that of starving people in famine-stricken regions where nutrition and survival outweigh all secondary motives, other species be damned.

Whales were hunted close to extinction, but now, in civilized societies, other (economic) values may surpass that of the meat. For many species the modern economy is both a curse and a blessing. Efficient killing and ruthless development threaten their existence, but the concomitant affluence offers hope for better coexistence. Diversified values are personal and more difficult to quantify,[13] but our economy relies crucially on them. Whale watching, for example, can now be more lucrative than whale-hunting. Gradually our economy will be less dependent on basic needs that are typically featured on the supply–demand laws of mainstream textbooks, and instead be more dependent on products and services catering to our ever more diversified wants.

The affluent can also afford significant deviation from narrow self-interest. For example, Bill Gates and Warren Buffett—among the world's richest—have pledged large portions of their fortunes to altruistic causes which they deem worthy, instead of to their heirs. Such deviation is not known in Nature. These aims are now much broader than survival and the number of offspring.

Secondary motives might often conflict with the primary motive, and this may not be penalised by market forces. In Nature, those less selfish will be doomed. Natural Selection allows little room for non-selfish deviants, whereas informational selection is more clement.

Most mainstream economists readily admit that man is more complex than *homo economicus*, but in their research they still focus on the primary motive. But why bother with idiosyncratic deviants? If the deviations were indeed random and irrelevant for the economy, then they would be justified in focusing on the primary motive only and leaving aberrations and anecdotes to the "lesser" social sciences.

However, the main theme of this chapter, and indeed of the whole book, is that these deviants are not random. They provide the inexhaustible source of our wants and skills that power economic growth.

8.2 Inverted pyramid

In this section we advance an even stronger prediction that the mainstay of the future economy will be based on secondary motives. Some may retort that the economy is about money, so how can we say that secondary motives can rival the primary motive? To understand this seeming paradox, it is instructive to examine what constitutes the current economy and how it will evolve.

Centuries ago an economy was measured by how much crop could be harvested, then by how much steel might be produced. In the twenty-first century, most of our wants shift from survival to secondary needs that in the aggregate take an ever-larger portion of the household budget. At the same time, more and more people work in jobs with skills far more diverse than the hard labor of former times. Man is on both sides of demand and supply; for example, an engineer is a generalist consumer as well as a professional specialist. Here we focus on man's wants, and in Section 8.3 we dissect his skills.

Newer and more diversified consumer products cater more to our peripheral wants than to our vital survival needs. Peripheral wants are great in number but low in frequency, and many of them we know only implicitly (Sections 2.5 and 3.2). Each of them may seem trifling and negligible, and if any of them is missing it is not a big deal for our lives; but taken together they represent a growing fraction of the total economic output. In fact, they will take the lion's share of the future economy.

The most frequently needed items are on our explicit to-do list, but peripheral wants are much less obvious. For each of our peripheral wants, consider a hobby of marginal significance that we may even forget unless being reminded. If we were to provide the skills for that hobby as a profession, it would require our full attention and we would face stiff competition.

Consumers cannot precisely articulate their preferences, and often they do not mean what they say. This poses a problem for businesses as they cannot literally follow the motto "do whatever my customers say." For example, Ford[14] once considered adding an extra door to its then popular pickup trucks, and asked their clients whether they wanted this innovation. The answer was "no." General Motors, however, instead of asking questions, studied people's habit in using pickup trucks and concluded that an extra door would be popular. The new design turned out to be a success. Ford belatedly added it, but the delay caused an estimated loss of $1 billion. Consumers were unable to articulate their implicit wants, and only in the right context could they understand what they really wanted. Shrewd entrepreneurs can predict consumers' wants to an extent, and new technology and new business models can detect useful signals in seemingly erratic behavior. They do not merely listen what customers say, but from observations they decode their true intentions.[15]

Secondary motives are behind man's numerous peripheral wants. For example, the additional features of upgrading iPhone 8 to iPhone x are probably not that vital for many poeple. But it is another story totally for Apple, as without constant renewal its dominance, and indeed its survival, are not guaranteed. For each of the new features, Apple engineers work with the utmost attention to push the limits of technology. Although each new feature can be of negligible importance to consumers, trifling innovations collectively constitute the cornerstone of our economy.

Inspired by Abraham Maslow's theory of hierarchical needs,[16] we consider secondary wants as hierarchical. Differing from Maslow's broad needs of personal well-being,[17] however, here we focus only on those needs having the potential to become commercial wants.

Peripheral wants, like idiosyncratic hobbies, are more personalized and hence more diversified across a population. We all have the same basic needs—food and shelter—but our higher needs can differ as they are often cultural, recreational, and intellectual pursuits.

Everyone has a great number of idiosyncratic, low-frequency wants, and needs. In other words, implicit needs are still in our head but are dormant, and our raw brain-power cannot afford to keep them perpetually active in our to-do list.

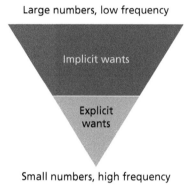

An inverted pyramid.

We propose an inverted pyramid (as in the Figure) to represent hierarchical wants. To emphasize that higher wants tend to be more diversified, it is convenient to place Maslow's pyramid of human hierarchical needs on its head. This inverted pyramid emphasizes that basic needs are fewer in number, whereas higher wants are much more numerous. Note that basic needs are more frequent, and higher wants are less frequent.

Maslow[18] suggested that "a human being is a wanting animal and rarely reaches a state of complete satisfaction except for a short time." As emphasized previously (Sections 1.5 and 6.1), our wants not only shift from the old to the new, but also from the few to the many. Man's needs become personalized and diversified when he is well fed and sheltered, and affluence can also make man's values diversify.

The wide opening of the pyramid's upper parts suggests that secondary wants are more diversified, idiosyncratic, and personal. Businesses' offers in the aggregate are as diversified as consumers' secondary wants, but each individual firm must retain a narrow focus.

Diversification is on two distinct levels: personal and systemic. Everyone has his own inverted pyramid representing his diversified wants; your pyramid and mine differ in the upper parts. In the aggregate over the whole population, we have a collective pyramid with a much wider upwards opening, as systemic diversity is much more extensive. Subsequently we shall in general refer to the economy-wide pyramid, unless mentioned otherwise.

Contrary to the well-known bottom, the upper parts of the pyramid are less well known. Therefore, consumers' secondary wants tend to be mediated by matchmakers—among them our Personal Assistant. There is no end at the top, suggesting further extension and diversification. The future economy, while consolidating the entire pyramid, would expand upwards without end.

Secondary wants are too numerous to always keep in our active memory, and on them our infocap is ever weaker as we dig deeper into our cognitive depths. The difficulty is further compounded by the fact that while the basic needs remain stable, the higher wants are fast changing.

The expanding secondary wants often remain implicit, but with a suitable stimulus they may be awakened. Matchmakers and PA can play a vital role in selectively converting our implicit wants into explicit wants. We delegate more and more of the hard work to PA and other matchmakers.

Many of our wants and skills are implicit and hidden,[19] and the challenge is in unearthing and exploiting them. Only a small fraction of our wants and skills ends up in the supply and demand that register in economic transactions; and yet the hidden part plays the most important role, as it can provide inexhaustible sources for powering economic expansion.

The future economy will see more and more of its GDP arising from catering to our secondary wants on the upper parts of the inverted pyramid, as in the aggregate they are more important.

8.3 Selected secondary motives convert to the primary motive

In Section 8.2 we analyzed the secondary motives behind man's wants. In this section we examine man's skills which ultimately power businesses' supply.

We may think that wants may be frivolous but professional skills must be driven by the primary motive; but this turns out not to be the case. There is a huge pool of personal skills that can power the economy, and most of our skills are also from secondary motives.

During the Stone Age, man's primary means of survival was physical force. In the economy of Adam Smith's time they were land, capital, and labor. In the twenty-first-century economy, man pursues numerous seemingly frivolous activities, and there are many more diversified ways to make a living.

Skills are also hierarchical and can be represented on a similar inverted pyramid.[20] For example, higher skills are personal, and their number is much larger than that of hard labor. Our basic skills have a high communality, and higher skills are more diversified from person to person. There is, however, a crucial difference: the inverted pyramid is a correct representation for a single consumer's wants, but this might not be true for his skills. An individual might have many skills, but those that are of actual use in the economy are very few. On the other hand, aggregate skills in the economy form the inverted pyramid with a diverging upper part.

There is a subtle difference between wants diversification and skills diversification. The former is spontaneous, but the latter is deliberate and hard sought, and its path is full of failures and dead-ends. Fundamental asymmetry (Section 1.4) also bears on individual's wants and skills. A personal want can easily be saturated, but when we develop a skill it can serve many people and may play a significant role in our life.

Everyone has many wants—realistic or not—as well as many skills. We hope to be rewarded for a few of our skills to pay for our numerous wants. The problem is that among our skills, whether from childhood dreams or earnestly pursued passions, only a very few of them are selected by the economy, due to competition. By contrast, hardly any of our needs, however frivolous, will be ignored by the economy.

So, how are man's skills selected by the economy? Informational selection on man's skills acts in two stages: first an individual must mentally select from what he knows and focus on a few; then society must find, compare, and evaluate his skills from among many competing individuals. The outside world in fact knows only a tiny fraction of your skills; moreover, you have far more implicit skills than you know of yourself— and for those you do know, it is still far from certain that you can articulate them clearly. Therefore, man knows more than he can tell, and what he can tell is still much more than can be found useful in the economy.

On the surface it may appear that our skills are driven by the primary motive, as making money is a key part of it. We might be negligent in forgetting some items on our list of wants, but for what we do for a living we must know better than anyone else. And yet knowledge can still derive from secondary motives at the starting phases. It is a long process from a fleeting thought to an absorbing passion.

Teaching children to be rational and to grab every opportunity for their own gains may not necessarily be more productive than letting them pursue what their fancy will lead to. Secondary motives play a far more important role than is recognized. Over time, all of the mainstay economy can be regarded as the result of hesitant lead that happened to be selected, amplified, and cumulated.

Children are born curious tinkerers; adults are burdened by daily toil. Many of our secondary motives are suppressed while growing up, but some endure and expand further.[21] Some secondary motives represent personal values and can be goals per se independent of, or in conflict with, the primary motive.[22] Man's secondary motives present a complex subject, but as far as the purpose of this book is concerned, a small fraction of them are selected to power the economy.

Our knowledge can be similarly represented as an inverted pyramid. In fact, the knowledge pyramid is closely related to that of consumers' wants, as well as to that of people's skills.

The large upper parts represent now implicit or hidden knowledge that a person cannot articulate or recall, but which may become explicit upon suitable stimuli. The middle parts are personal knowledge that is explicitly known and recallable, but is not yet revealed to the outside world. Finally, the bottom part is the shared common knowledge. The inverted pyramid for knowledge emphasizes the fact that the upper parts are much larger than the lower parts, and implicit knowledge is limitless, depending on how we exploit it.

Most of man's knowledge remains implicit. Explicit knowledge is what we can clearly identify and recall when in need. For example, implicit memory hides in our head but may[23] emerge in suitable contexts; scholars, in fact, consider our memory situational. The recall function draws from explicit memory, and the recognition function from implicit memory; we recognize much more than what we can recall and recite.

Take your most favorite book and close it. You can hardly recite it, but if shown a passage you may recognize the words in detail. Consider one of your favorite childhood songs. If the music starts to flow you may remember each word of the song. IT can help reveal and harvest (Section 2.5) the vast pool of our implicit wants. Polanyi's pioneering work[24] emphasizes the implicit dimension in personal knowledge, and here we focus on the role of implicit knowledge in the economy.

Skills that cannot be articulated are difficult to transmit; for example, an apprentice must observe his master closely, sometimes for years, to learn the trade. Yet even when the master genuinely wants to transmit his expertise, he cannot articulate all the subtle details.[25] This transmission difficulty also arises between cooperating colleagues, children and parents, and pupils and teachers. However, this difficulty can be a boon for innovators, as it provides a welcome delay which allows a temporary edge over the imitators.

Mental selection was first studied by Thorstein Veblen. In his book *The Instinct of Workmanship*[26] he states that when people are not particularly pressured, their idle curiosity may lead to creativity. Experts might reduce the number of trials and errors, but selection cannot be entirely avoided. From a fleeting thought to a successful innovation, the steps do not follow a chain of logical events that scholars try hard to unearth. Hume's *A Treatise of Human Nature* reminds us that untold large numbers of discarded ideas and variants also play an important role in every success.

On the level of the economy, man's talents must be uncovered and appreciated as such. Informational selection for useful secondary deviants operates on two levels: first, mental selection, and then selection through various steps in the economy. Workers choose alternatives during production; they are chosen by managers, who in turn are selected by someone higher up.

For example, an IBM report examines how inventors invent,[27] to ascertain what circumstances are favorable for creative sparks. Engineers and scientists often follow their instincts and curiosity without strictly toeing the company's line of profit imperatives. Creativity knows no corporate boundaries.

Thus, an IBM engineer might register a chemical component patent, or a DuPont scientist might invent a biomedicine; but neither may be directly useful for their respective companies. A matchmaking company, yet2.com, aims to find suitable matches for many of the dormant patents locked up in corporate safes. It claims that as many as 90% of all patents in use have been created by someone else. Creativity responds to neither monetary incentives nor command.

Abraham Maslow, in *Motivation and Personality*, states that creative minds need a foggy screen.[28] Throughout this book we disapprove of the weakness of informational selection and seek ways to improve it. Here we seem to advocate for milder selection criteria, but being milder is not equivalent to being weaker. If the tasks are complex the selection criteria should not be too rigid, leaving sufficient room for the unexpected. Google, for example, allowed its engineers one day per week to freely pursue their favorite projects. Good education also aims to instigate analytic power to cope with the unknown rather than to provide a fixed tool box.

The saying "necessity is the mother of invention" does not contradict the assertion that creativity cannot be commanded. Outside selection pressure can act on deviants a posteriori, but it can do little at their genesis. Artists during the Renaissance created a great cultural heritage that is still arguably unrivaled—not that people before or after are less talented, but during that time the emphasis on art enabled promising artists to be recognized.

Novelty relies as much on a priori genesis as on posteriori selection.[29] People's appreciation (selection) has strong feedback to the inventor, who may not always be aware of the full potential of his own creation. If an innovation is appraised in timely manner, the inventor will be further galvanized into doing more, and better. The opposite is also true if it is not recognized, the inventor's initial creativity might fade. Talent is also perishable.

Information Technology differs from traditional technology in that that there are far more unintended uses beyond the original intention. Computers were designed for serious computing,[30] but today they are predominantly used for a great many seemingly frivolous purposes far beyond the inventors' imagination. Alphanet was built to connect US government computers. Can we really draw up a list of the uses of the Internet today? Secondary motives emerging on the Internet are not random, as shrewd entrepreneurs can find exploitable patterns in apparently unpredictable man. Informational selection in the Internet age relies on a much larger number of secondary motives and deviants, but it picks up only a small fraction of them. IDOL (Section 2.3) is without economic incentives, and yet it can have a powerful impact on the economy. Although most deviants do not make (narrow) economic sense, they allow the economy to achieve far more than the primary motive alone.

Sergey Brin and Larry Page did not set out to become billionaires. They first realized that they had an edge in doing something better than anyone else, and only later, after Google's initial success, did they find a way to generate revenue. What served the purpose may not have originated from purposeful actions.

Mark Granovetter has analyzed how entrepreneurs used existent social ties for business aims.[31] These ties were not created for profit proposes, but some of them were found to be useful later. It is farfetched to say that friend-making is for maximizing profit, but those with many friends may indeed obatin help in unexpected circumstances.

In his well-known work, *Strength of Weak Links*, Granovetter found that the social links in which we have invested the least may turn out to be the most valuable with life-changing potential.[32] Gary Becker, on the other hand, considered that most human activities, such as church-going or child-bearing, maximize utilities.[33] With the advent of global social networking sites, people can access a much larger number of weak links than those in Granovetter's study, and these chancy encounters can lead to more business opportunities.[34]

8.4 Formable man

Besides merely selecting from the existent secondary deviants, some of them can be deliberately cultivated. Therefore, informational selection differs from Natural Selection in one more way: the raw matter (deviants) can be created purposefully.

This assertion pushes economic constraints further; if man's propensities can be cultivated, then they may be intentionally channeled to gainful ends. Throughout this book we regard wants, skills, and values as changing, and in this section we focus on intentional changes over time.[35]

This writer had a firsthand experience during a holiday trip. Upon leaving the Maldives, a guide suggested that tourists should take away the rubbish of the day on the flight back to Europe, to protect the island's fragile environment. But a fellow traveler—an engineer by profession—retorted that by carbon dioxide emission standards it would be more rational to throw the garbage in the ocean than to fly it thousands of miles to Europe. These calculations may be factually correct, and by strictly following Cost Benefit Analysis (CBA) the ecological gesture might seem insanely irrational.[36] Most travelers nevertheless heeded the guide's suggestion, and even the children took bags of rubbish onto the flight. What the engineer missed, however, was the bigger picture. For the tourists who fly with rubbish there is a conscience-forming role that CBA cannot compute, and such a gesture might have lasting effects on the future behavior of tourists. The chances are that they will be less likely to degrade pristine Nature with rubbish, and their next purchase of a car or a house, the environment may be a more important factor.

An article concerning William Lindsay—a British retired teacher— reported that during a visit to China he was dismayed at the unsightly rubbish left by tourists on the Great Wall near Beijing.[37] He therefore decided to return to China and pick up the rubbish. Anyone well versed in CBA would say that his additional journey from the UK to Beijing would cost far more (in money or emissions) than his intended result, but what appears irrational may turn out to be rational on the bigger picture. Lindsay's action found unexpected support. School pupils soon noticed the unusual visitor and spontaneously joined him in picking up beer cans and cigarette butts. Then, a local newspaper published a report, schools organized rubbish-collecting excursions, and citizens, farmers, and tourists participated in the clean-up. By the time William Lindsay returned to the UK the international press had reported the story, and later, Queen Elizabeth II bestowed on him the Order of the British Empire. A similar story concerns the Japanes retired mountaineer Ken Noguchi, who decided to clean up Mount Fuji.[38]

It may be said that people on the fringes often do crazy things, but businesses can also defy narrow rationality. For example, the shoemaker Zappos offered an amazing return policy of 365 days with two-way free shipping. Experts considered such gestures to be commercially foolish; and yet the business succeeded on the bigger picture. Despite some opportunists taking advantage of the over-generous policy, many customers saw it as a gesture of trust and confidence in its own product.

Ordinary people tend to heed their instincts and seem to fail on the narrow rationality criteria, and yet this may make better sense on the bigger picture. Michelle Obama with her daughters growing vegetables in the White House garden might not be economically rational, as was pointed out by the pesticide lobby in an open letter to her.[39] The lobby was afraid that if the First Lady promoted natural produce without pesticides it would set a "bad example" for the nation.

Google has the Café 150, with food sourced from farms no further than 150 miles distant. This can also be regarded as gravity-defying, as the "eat-local policy" has clear costs but with obscure benefits. Solar Impulse designers spent lavishly on a fragile aircraft that accomplished the first round of world flight on solar power alone. Many tons of carbon dioxide were emitted during its construction, and its practical utility is null, all of which can make sense only with larger scopes.

Bigger pictures necessarily touch upon values that cannot be quantified, and the priceless has no place in CBA theory or in mainstream textbooks. The example of Google's Café 150 does not show that rich kids waste money—they still count pennies when negotiating deals—but that affluence allows them to pursue less quantifiable values.

Habit-forming can also reinforce the primary motive. Wal-Mart used to ask its executives to collect free pens during their stays at hotels.[40] The multi-national company must, of course, pay for hotel accommodation, so why bother with hoarding pens? Is Wal-Mart over-greedy or shortsighted? It is probable that it intended to imbue a culture of frugality: if an executive makes the conscious effort of collecting hotel pens, in his other dealings he will be parsimonious.

Philosophers such as Karl Popper believe that theories adapt to reality until they fit. Instead, incorrect theories may change people rather than the other way around, and this is consistent with the views of Thomas Kuhn and George Soros. Economics education inadvertently makes students downplay bigger pictures and other values that cannot be counted, and urges them to think like an economist[41]—following the narrow rationality doctrine and being subject to fewer spontaneous follies than ordinary folk.[42]

Man can be both baser and nobler than he is, and he may become either a saint or a monster.[43] Teenagers may dream of being a doctor with the noble aim of saving life, but "questionable doctors" (Section 2.2) did not start out to defraud patients for profit; bad institution rules led to their derailment. No-one is born a villain or hero; it depends on how people are embedded in society. Man should not be viewed as an opportunist ready to grab any chance for his selfish gains at the expense of others, nor as an angel depicted by romantic thinkers.

A too rosy view of man ignores his baser side, and a too pessimistic view fails to recognize his upside potential.[44] It is undeniable that there are brave individuals willing to sacrifice their lives for altruist causes, but most us are still subject to the gravity pull: our primary motive. Societies based on noble ideals solely risk exploitation and abuse of goodwill. Employees who are entirely sustained by the love of the profession cannot go far without the primary motive. Even heroes still need compensation such as praise or peer esteem, and a society cannot be sustainable if it relies on heroism alone.

Ants exhibit coherence toward a common goal; rulers of totalitarian regimes dream of having their subjects achieve ant-like harmony, and sometimes they might succeed by coercion. Repression of secondary motives inadvertently damages the primary motive: forcing is not forming. For example, North Korea may impose a high level of harmony peerless in the world, with tens of thousands of citizens dance in festive unison for their rulers; but an indirect result is the country's failing economy.

Human intelligence makes it difficult to band people together unquestioningly like ants. People's interests partially conflict and partially overlap among them. Man's baser and nobler potentials are unlimited, and outside selection plays an important role: for the short term, restraining; for the long term, habit-forming.

Forming habits may be intentional but cannot be deterministic; innate talent is still important, but the debate of nature vs. nurture may continue forever without a clear winner. For a budding golf or tennis talent, a big budget will help to propel him to the top, but the same budget would be a waste if invested on a random candidate. Social capital, trust, and habits are built up slowly, but they can be destroyed within a short time. Goodwill can emerge spontaneously, but it is also perishable, and society can either amplify or degrade it. People cannot defy gravity indefinitely, but some habits may result in lasting changes in behavioral patterns.

9

Evolving Economy

This chapter further develops economic evolution powered by information selection. We observe that in economic evolution, institutions play an important role, and many new issues not encountered in Natural Selection appear. Economic evolution often does not take place by pure selection (choosing or eliminating), but by building an elaborate institution. For giant institutions such as global platforms, the numbers from which to choose are probably very few, but selection is nevertheless always present, as it operates on the smaller units within.

9.1 Communication costs going down

An important barrier against informational selection is communication costs, which steadily trend downwards and include collecting, transmitting, ranking, sorting, filtering, and analyzing. The reduction in cost impacts on the economy not only by incremental changes but also by occasional disruptive upheavals. Established business models can become obsolete and new models emerge.

For example, traditional advertisements in our mailbox cost a few cents each to print and deliver, so their quality is tolerable, and local businesses target their neighborhoods with relevant promotions. Spams in our inbox cost almost nothing, hence they are less relevant. One might think that the cost of a few cents is negligible, but it is far different from being zero. Electronic messages cut distribution costs dramatically, and spams become a new plague that accounts for a significant portion of Internet traffic.

Alan Kay—an engineer at Xerox's Palo Alto Research Center in the 1970s—has remarked that reductions in the cost of hardware led to an increase in "wasting transistors."[1] Half a century ago, every computer had specific purposes, but as the price per transistor has constantly dropped, computers now carry out frivolous functions that are so numerous that no-one can list them all.

Reductions in communication costs can also give rise to new risks. Book-lending among people is a common practice, which is why we have libraries. If dozens of readers can read a single book, publishers rarely complain, but although physical barriers made lending a book to thousands or millions people impractical, it is an entirely different matter for e-books.

Such barriers are hard to notice until when they are much reduced; an altruist buyer of a single copy of a book or music disc can make it available to the whole world. Books and music were once bundled to physical carriers, but now content can spread without bounds.

There are many new ways of collecting information.[2] For example, RFID, biosensors, and myriads of other devices can capture written, audio, visual, and chemical signals. This inevitably leads businesses to rethink their revenue models.

With the ease of collection and storage a massive amount of data is available, but our capabilities of data analysis are a severe bottleneck, as mentioned previously in the context of Personal Assistant. The rise of social networks such as Twitter and Facebook is not due to hardware advances or software development; rather, they are inventions on a higher level, simultaneously involving technology, economy, and society. The limiting factor for further innovation is our imagination in this unfamiliar interdisciplinary space, and many entrepreneurs struggle to combine IT tools with human needs and skills that are often seemingly frivolous and marginal.

Reducing communication costs and improving data analysis capabilities will further unleash man's new wants and skills in ways that we cannot fully imagine now. Established business models will be upset, and new business models will bring unexpected opportunities and risks. Indeed, reductions in communication costs will reveal a strange new world.

In physics, the further we dissect matter the more simple a world appears. From mountains and elephants to molecules, and on to atoms and quarks—the sub-atomic world is the simplest one. While physical matter appears more simple the deeper we probe, however, in socioeconomic studies we find exactly the opposite; upon closer examination, economic problems become ever more complex.

9.2 To unleash

Throughout this book we have shown that diversified wants can manifest in large numbers. In this section we instead focus on man's capabilities; reduction in communication costs also makes productive force easier to exploit, in unexpected ways and from unlikely quarters.

Platform matchmakers (Section 2.2) have enabled much of the current "sharing economy." For example, Airbnb and Uber have revealed previously untapped productive potential in ordinary people. The traditional separation line between producers and consumers is blurred and transcended.

Some previously unsuspected resources can now be exploited: consumers can participate in product design, and "production" loses its traditional meaning. In the past, businesses exploited their customers' weaknesses for profit, but now they solicit their help in establishing new products; Kevin Kelly[3], in *New Rules for the New Economy*, advocates that smart consumers are what they really need.

The newly awakened activism is prone to controversy. For example, Pervertedjustice.com keeps a watchful eye on paedophiles, and its members can expose even those who have not yet committed any crimes but are already engaged in lurid online communication with children. Although such systems do not and cannot replace the police, making offenders ashamed is often a better deterrent than other legal options. The court of public opinion has a new meaning, and criminals must now reckon with citizens' online eye. Overzealous citizens have sometime made mistakes by misidentifying innocent people; but the police also make mistakes, and yet no-one calls for them to be disbanded. In Chapter 2 we mentioned that patients might wrongly complain about a treatment, but doctors sometimes amputate wrong limbs.[4]

The music industry struggles for new revenue models as the costs of recording instruments and physical discs are in free fall. Major costs for music production arise from talent selection and promotion. Music fans should be treated neither as passive consumers nor potential thieves, as they can be enlisted as a productive force. When enticed to select music, fans do it with pleasure. Like bees sipping honey from flowers, while enjoying it they inadvertently provide a valuable service. The public may be better judges than the executives of big music labels.

Michael Lewis' book *Next* reported that with music fans selecting music from several hundred thousand artists[5] it is possible to tap into the vast infocap of fans for co-production. Kickstarter aims to disseminate creative work earlier and more easily by allowing investors (most of them are also fans) and artists share the risks and successes. Rather than belatedly generating megafortunes for the very few, it provides a decent income for the many.

Partial appropriation does not spell doom for content owners, but they need to change their mindset and revenue partition models. They can learn from search engines where the main service is provided for free, but many fringe benefits can pay handsomely. Some musicians, for example, already provide online music for free and derive a good income from live concerts.

As communication costs reduce, content owners can either play defensive games or seek new pies. Some set up artificial impedimantes to hinder free flow. For defensive examples, Amazon ensures that you cannot read more than a few sample pages of a book, Google Books might show many pages but with occasional omissions, and Salon.com will not allow further browsing without first clicking on the advertisements.

Hindrance tactics sometimes go too far. For example, at a Prince concert, fans had their cameras, camcorders, and smartphone batteries confiscated at the entrance after a body search[6]—but the organizers should know that those who record such events might be the most devout fans, and that photographs and films can be the best viral marketing. Scholars also offer defensive advices: for example, Harvard law professor Terry Fisher advocates taxing discs.[7] Content owners appeal to the government, urging it to use taxpayers' money to sue fans.

The government could play a more proactive role. It now often makes its employees care more in playing safe than in taking initiatives; they face little incentive on the upside but huge risks on the downside. The government has a unique institutional role that cannot be replaced, but it often sides with the incumbent; in an overzealous bottom-plugging role it inadvertently inhibits enabling the top, and legislation often limits rather than fosters innovation.[8]

Institutional rules are important for both unleashing the top and plugging the bottom. The top represents the noble side and the bottom the base side. Enabling institutions (such as Airbnb and Uber) are different from constraining institutions. Unleashing the top is harder than stopping the bottom: the top goes into unknown space where man's motivations, habits, social connections, and IT tools have countless combinations, while the bottom is relatively easier to plug, as all we need do is to forbid known illegal conduct.

Unleashing the endless potential in the economy will rely heavily on the enabling institutions. Whereas constraining institutions are now prevalently in the government, most enabling institutions are from entrepreneurs seeking win–win–win outcomes.

For example, connecting people via reputation institutions limits their baser side, whereas the nobler side is also enhanced. Better institutions bring out a better man, and a better man will be more likely to further strengthen better societal institutions. It is a slow force under which man and society coevolve and redefine each other in a reflexive, virtuous cycle.

IT together with institutional rules can lead to an explosive unleashing of creativity. Hardware technology is already one step ahead of its full use, and man's potential has never been revealed so much as at present. Man and economy coevolve, and institutions play the most important role in unleashing future unlimited potential.

After millions of years of evolution, humans have now suddenly become activist and inventive—but the potential has been always there, waiting to be awakened. Cognitive power is at about the same level as when man needed to outsmart beasts in the Stone Age. If man can achieve infinitely more complex tasks today, it is due to the coevolution of man and the economy.

Over the ages, bees have been building nests, and crocodiles and sharks have not changed over millions of years—and yet the economy will never settle, nor will man. Collectively, humans are far more powerful than other creatures, due to the coevolution. If man is wise enough to not self-destruct, the potential for further development is indeed beyond imagination.

9.3 Bigger picture and longer view

In this section we emphasize that enhanced informational selection not only promotes economic growth, but also provides a bigger picture of human welfare that is sometimes at odds with narrow and short-term economic imperatives. Economic development supposedly overlaps with man's welfare, but often there remain serious conflicts.

For example, prevention and healthy lifestyles can avoid many painful and costly treatments, GDP be damned. On the other hand, if the fast food industry makes people obese and the pharmaceutical industry sells them slimming pills, GDP may grow but human welfare (non-priced) suffers.

Healthcare does not face a fixed number of diseases, as many of them are preventable. With severe information asymmetry we cannot rely on market mechanisms alone. Pure economic incentives will not produce better results, as revenue pressures often outweigh the welfare of patients.

The vaccine preventing cervical cancer costs only a few hundred dollars per person, but its huge welfare hardly registers in GDP. For example, Novartis's Gleevec (or Glivec in some markets) was hailed as a wonder drug in the management of leukemia, and was priced about $100,000 per year for keeping a patient alive (but not cured). Gleevec contributed much to GDP, but the improvement of welfare may be less than that of the vaccine.

It is common practice that the pharmaceutical industry provides many gifts and perks for doctors.[9] But there should be higher-level prevention, which is even more important: disallowing the practice of doctors' income coupled to drugs and treatments which they prescribe[10] will reduce drug overuse at the expense of GDP. When incentive policies go wrong we may observe strange phenomena. The US surgeon and author Atul Gawande, for example has reported that McAllen county hospital carried out more MRI scans than the whole of Canada.[11]

What is good in one context may be questionable in relation to the larger picture, and new pies might not always be good for a sustainable economy. Consider the example of embedding many trendy electronic gadgets into refrigerators. Traditional refrigerators have a lifespan of about nine years, and some firms hoped that the newer machines full of gadgets would have the expected lifetime shortened to three years. In an article in *Businessweek* this was hailed as an important innovation, speeding up the consumption cycle of refrigerators and considerably boosting corporate profits and GDP.[12] In his book *Free*, Chris Anderson (author of *The Long Tail*) praises the issuing of free newspapers, but he does not mention the massive amount of waste paper that is generated.[13]

The mainstream picture can be notoriously narrow. The economist Frank Hahn has said that "we need not worry about exhaustible resources because they will always have prices which ensure their proper use."[14] On the other hand, according to David Hillis, a biologist at the University of Texas, about 90% of all living species are not yet described and much less priced.[15] If one is convinced of the mainstream doctrine, the unknown will be doomed.

This doctrine may seem ridiculous, but it resurfaces regularly in many disguises. The current pricing scheme uses the cost of extraction as the measure of a resource's value.[16] But what is the value of something which can never be replaced? This becomes even more difficult to account for when it is only future generations who will feel the full effects of the depletion. Any intervention will benefit some and hurt others, and our theory leaves many intriguing open questions concerning who should exercise them, the criteria, and so on. In the bigger picture there are usually many winners and also a few losers, and it is rare that everyone wins.

For another example, Cargill, a big US food company, campaigns[17] for more salt consumption by Americans to "make sure you have plenty of salt in your kitchen at all times." On the other hand, experts estimate that cutting the already high salt intake would save 150,000 American lives each year. Salt is cheap for most people, and to double the net worth of a few salt millionaires we need to swallow several millions of tons more of salt. While on the bigger picture of human welfare the choice appears evident, the salt producers could not care less. Gains for the minority on salt, pesticides, and unnecessary surgery harm human welfare but enhance GDP.

There is no ideal intervention that hurts no-one,[18] and society must deliberate on whether to protect the salt or cigarette industries' profits balanced against the health of consumers.[19] Bigger pictures and longer views can be considered effectively only by higher institutions, while the parties which are directly concerned are often powerless to counter the abusive practices harming the vast majority's well-being.

For longer views, people might accept temporary losses in order to achieve better gains in the future. For example, a renewable-energy policy might oblige us to accept the extra costs in the short term; but we might refrain from exploiting pristine Nature, hence incurring a price for the economy, for the unquantifiable benefits of future generations. Renewable energy is often more expensive than fossil energy, and over time the economy of scales can reverse the cost differential. This type of loss-leader requires the bigger pictures and longer views, which would not spontaneously happen without deliberate policy designs.

Man's multiple non-economic values are often in conflict with economic development in the short term, but they boost it in the long term. Institutions can help people to realize the long-term welfare and refrain from the temptation of short-term gains. Businesses are often thought to be opposed to environmental protection in order to produce short-term profits, but this belief has proven obsolete, and entrepreneurs may have bigger pictures and longer views than do many mainstream scholars and politicians.

9.4 Economic evolution

It is often suggested that in many respects, economics is like biology. Charles Darwin benefited from the insight of classical economists such as Adam Smith and Thomas Malthus in formulating his theory of evolution. Natural Selection is about how we came to be here, whereas informational selection concerns where to go from here. In this section we compare the selection mechanism of the economy to that of Nature.

At the end of the last Ice Age, giant mammoths were extirpated by natural elements. Now, gazelles in the Serengeti, for example, might eventually die out if they cannot run sufficiently fast in order to escape predators. In our modern economy, agents do not face the same harsh selection pressure, as merits and rewards are not tightly correlated and the motto "survival of the fittest" is less true in the economy than in Nature.

Unlike in a Robinson Crusoe economy, we do not make our living directly from Nature exclusively, as most people work inside the vast transaction zone between production and consumption. We rarely make our own cloth, food, or dwelling, and we trade our skills for our wants in the economy. Since our performance is judged by fallible humans, selection strength is weaker than that of Nature.

Selection naturally leads to evolution. It is instructive to further compare the evolution driven by informational selection to that by Natural Selection. Brian Arthur, a Santa Fe economist, has noted that evolutionary pace is much faster for man-made processes than for those in Nature (a ratio of 10 million to 1).[20] Biological evolution is hardly discernible over a few centuries, and for practical purposes, species can be regarded as stable. A few years can be a long time during which our economy might experience dramatic shifts.

The analogy between sexual selection[21] and informational selection is relevant here. In biology, higher creatures such as mammals—the dominant form of life on Earth—adopt sexual reproduction rather than asexual reproduction. Sexual selection can be said to be on a higher level. It is the earliest form of informational selection, but is more vulnerable to cheating. It also demands adequate infocap for selecting quality mates. It is difficult for a peahen to determine which peacock is more fit, since she cannot procure a physical examination of her purchasers. Imperfect infocap in sexual selection gives rise to deception, and only credible, costly handicaps can convey the true fitness. The conspicuous display of ornaments is a huge waste of resources that might have been invested in physical survival.

Higher economy is paradoxically less perfect than lower economy. The barter economy is closer to perfection than is the modern economy. For example, entrepreneurs of information economy can appropriate only a small fraction of the value they create, whereas in the primitive economy, appropriation can be considered almost full.

This lack of perfection can be a good thing, and economic evolution is sufficient flexible to accommodate human values that are far broader than economic efficiency. We might deliberately choose a direction that might not aim for the largest GDP or fastest growth, but it is a balance between these economic imperatives with the other values befitting our prevailing enlightenment level.

In comparison to Darwinian Natural Selection, informational selection has a unique feature: foresight and selection complementing each other. Foresight can bring the following advantages over Natural Selection, despite informational selection strength generally being weaker. First, foresight saves resources and time (for example, B2 represents foresight). Biological mutations are random, and economic alternatives are much less so. Man strives to plan future actions, and we do not need to try many obviously senseless choices. Informational selection here enjoys a major advantage over Natural Selection: design and plans can skip many random trials, and there is great saving in time and resources (for example, help at the B2 stage; Section 3.5). On the other hand, due to our limited capabilities and the complexities we face, foresight has a limited range, and selection at many stages will always play a crucial role.

Many scholars have embraced Lamarckism for explaining economic evolution, and have written off selection. But what appears as an inevitable sequence is, upon scrutiny, less so, as there are many hesitant intermediate steps and deviations, rarely known to outsiders and scholars, where selection plays a vital role.

Bats and radar can be regarded as results of Natural Selection and informational selection, respectively.[22] The former evolved without a plan and over a very long time, while the latter was developed rapidly and efficiently—but still there were many stages, and numerous alternatives must have been considered for betterment. Richard Dawkins has utilized the bat/radar analogy to make the point that a blind watchmaker could build a wonderful creature like the bat but that a human invention such as radar would follow a different path.

Second, foresight allows us to bear temporary losses for future gains. If intermediate steps are worse (Kevin Kelly calls this "devolution"[23]) than the starting point but the long-term gains can be better, foresight allows us to overcome the temporary barriers. Informational selection can cross barriers, so to speak, which Natural Selection cannot.

Entrepreneurs might reduce the prices of new products and subsume temporary extra costs to reap larger profits later. Nations might decide to switch to clean energy such as solar power and wind power, even though these are temporarily more expensive than fossil fuels, as huge economies of scale might later justify initially non-economic choices.

The dilemma of foresight vs. selection is simply an extension of the flexibility–commitment dichotomy in the context of evolution. As mentioned previously (Section 7.4), a compromise is needed, and neither extreme is desirable.

Foresight plays an important role but cannot replace selection. Sometime, institutions are also designed with a view ahead of current practices, but they often lag. This is a clear departure from the view of mainstream economics which posits that institutions move in synchrony with the economy.

When walking we move one leg at a time, and each leg alternatively takes the lead. Institutions and the economy evolve more in this way rather by how a car moves its wheels synchronically. Ideally, institutions should always move first; but this is often impossible, as economic evolution often breaks into new space, and people can lack vision. The lagging institutions are often obstacles (like many current copyright laws), and pressure from the practitioners usually builds up to force obsolete institutions to adapt.

"Dimensionality curse" is an expression often used to describe the space of possibilities being so large that it is impossible to assess. For hundreds of bakers and hairdressers in a big city we might be justified in saying "let the best survive." For global institutions, however, we have only a few alternatives to choose from, and merits and risks often take a long time to be recognized. Traditional market forces are ineffective for selecting the best institutions. Textbook supply–demand laws that might work for bread and ice-cream producers become irrelevant.

Economic evolution faces two additional perils. The first is that higher-economy institutions are big in size and small in number, and the space of possibilities is hence very large; and yet few can be actually tested prior to use. The second is that the pace of evolution is accelerating since the Industrial Revolution. More complex space and shorter time together make the selection of institutions particularly difficult, and to cope with the challenge we need to appeal to foresight and deliberation.

Economic evolution finally becomes independent of Darwinian Natural Selection, and we enter a new phase where stakes and risks cannot be greater. In Natural Selection almost nothing can go "wrong," while in informational selection, avoidable perils are many and we will be responsible for our own destiny.

9.5 The single boat

There is a feature unique to economic evolution that we shall call the *single boat* phenomenon. Recall that Natural Selection relies on copious alternatives to allow the fittest to survive.[24] In many areas of the modern economy the number of viable alternatives is rapidly shrinking, ultimately to a single choice.

The "take-it-or-leave-it" modus operandi of traditional selection worked well for Natural Selection and also for the early economy, but now it has become less effective. The mainstream doctrine that the market will always choose the best, might have been suitable for choosing bakeries among hundreds in a city, but would fail for finding the best stock market or social network where only a few remain.

The number of global matchmakers and institutions are bound to be small, and a single player might dominate (Section 2.2). Success often depends more on the first mover's advantage than merit. We rarely regard first useable products such as shoes as the final design, but a sufficiently good website such as Craigslist or eBay might dominate from the start and for a long time.

People on a single boat have the same basic fate to share. Throughout history, sages such as Sun Tze and Machiavelli affirmed that the winner cannot afford to be arrogant. Besides morality there is the additional reason of everyone's survival and prosperity. It is like a giant ultimatum game played on the global level; a disproportionally small spending in terror can deny welfare to everyone.

The world is not only more vulnerable to terrors, but many efficiency innovations in the economy can also unwittingly bring potential harm which transpires only much later. In the past, farmers could do whatever experiments they wanted, but today, bioengineers releasing genetically modified crop without regard to the long-term effects may alter the planet's ecobalance forever. For issues such as the environment, natural resource depletion, and Internet rules, we already face the single-boat reality.

The single-boat phenomenon brings both a curse and a blessing. The curse is that tinkering with a single copy may cause irreparable damages. For engineers, if their bridge fails they might learn to build it again; but bio-engineers might not have such a second chance.

Environment degradation bears on our future, and we do not have the luxury of trying many frivolous scenarios, as blunders may be irreversible. We lack the absolute standard to judge, and the best we can do is to take our cue from state-of-art knowledge and proceed with precaution wherever single boat risks are present.

The blessing is more subtle. Where competitors are few there is less need of the "must run or else" doctrine. Whether Facebook is the best social network or Amazon the best online shop, they are less scared than a business among hundreds of rivals. Global giants can afford to take other considerations into account, and can contemplate longer views and bigger pictures (though whether or not they do so is an entirely different question). Reducing alternatives to a single copy thus provides more strategic options for the sole survivor, who might accept temporary price cuts in view of long-term profitability, or pursue other values.

Almost all governments follow the maximal growth doctrine, as if it were the only way to a better life, but many critics say that this doctrine is at odds with human long-term welfare and that moderate growth offers many advantages.

Here we postulate the minimum impact principle: it is desirable to make the ratio of the economic growth vs. environmental impacts as large as possible. For example, for a given improvement having zero impact the ratio would be infinite. This principle differs from other radical proposals of slower or zero growth, since it still promotes growth whenever possible, but the economic results should be measured against its environmental impacts. Future business ethics can be based on such a ratio: for example, a business might boast of a high ratio if it achieves a large economic output with minimal impact. Such a ratio should be updated frequently, as new knowledge will show real impacts and hence the ratio will be a moving target. Different economies face different urgency—developed vs. developing economies, or the previous century vs. the current century—regarding growth vs. preservation.

Growth can take place in many dimensions, and economic growth need not necessarily involve drilling everywhere where oil might be found, as in pristine Alaska, to power ever bigger SUVs. If society can take a longer view, there are numerous eco-friendly pastimes to choose from.

The trend toward a single copy increases efficiency but makes the world unstable. Informational selection enters a new evolution phase as we approach a single copy: it will operate within. Selection will still operate on all levels except at the very top, while on lower levels and local scales we seem to be able to choose our actions with concomitant successes or failures. The successful seem to have followed a clever path; if only a single boat is left sailing we can no longer rely on luck only.

In the recent evolution literature there is a dangerous trend of placing blind confidence in emergence: whatever is needed, including institutions, will emerge by self-organized miracles. Without foresight, emergence can also lead to maladaptation, and there are still many covert disciples of Voltaire's Dr Pangloss advocating that our world is and will be the best of all worlds.

On some global issues the single-boat phenomenon is already here, and we are far from being certain of our path ahead. On the grandest scope that is mankind's destiny we sail into uncharted waters, and our greatest challenge is how to build a sustainable economy and society. The stakes cannot be higher.

In Darwinian evolution, *Homo habilis*—earliest man, who first evolved about 2.3 million years ago—survived for millions of years; but there is no guarantee of man's survival for another millennium. The coevolution of man and the economy has resulted in a society of ever-higher efficiency and ever-heightened instability, and both our creative as well as destructive powers are unprecedented. Without foresight, and by relying only on emergence, we can easily squander the opportunities that still remain.

10

Paradigm Shift

This chapter discusses a new paradigm. Mainstream economy is based on the optimal allocation paradigm, by which, typically any problem is reduced to an optimization goal with given constraints. We summarize the new market theory by noting that in many contexts, allocation is a process. Any movement towards a planned goal would redefine the constraints; therefore, we cannot reduce most economic problems to static optimization exercises. Instead, we must consider a dynamic process in which both allocation and creation of resources are presented as two inseparable pillars.

10.1 Methodology reorientation: curves vs. dots

Expensive products tend to be of higher quality, but it would be a stretch to say that there is a rigid causal relation. In the retail sector the observed correlation of price vs. quality is due to the consumers' collective selection; they must shun some and favor others, keeping producers and vendors on their toes.[1]

Imagine a marketing report showing many observations of price vs. quality for shoes; scholars might be tempted to draw a curve to highlight the correlation. There are many such curves in the economics literature, often hinting at causality among variables.

People try hard daily to beat averages. It is the diligent drivers—either discerning shoppers on the high street or investors on Wall Street—who sustain the observed correlation. By trying to beat it, the skilled and diligent can end up on the better side of the curve more often than by random chance (Section 7.2).

Those who ignore mainstream theories are the real drivers, and their selection effort is not accounted for in mainstream theories. The important question concerns whether the curve migh degrade or improve; studying the curve itself sheds no light, and the answer is among the dots around the curve. Such curves are aplenty in mainstream theories, especially in modern finance theory. For example, higher returns are observed to be correlated with higher risks for stocks. This hints at a correlation: higher returns might be obtained by taking more risks.[2] The Security Market Line (SML) in the Capital Asset Pricing Model (CAPM) is one of the best-known curves in economics. If investors were convinced of the SML curve, giving up due diligence, the index would degrade.

Mainstream finance theory fails to mention that its prescription would herd people in the same direction, and their combined weight can be considerable—contradicting its aim to educate all. Teaching the theory can undermine the existence of the curves: if more people were converted to be believers, the faster the correlation would degrade. This apparent correlation is not due to decades of mainstream praise, but *despite* it. The curves not only imply wrong causation but also induce indolence, and even those who can make a difference might be persuaded to give up trying.

Imagine that a certain disease has a 50% survival rate. If a doctor is only good at what is the equivalent of econometrics in medicine, his patients would end up on the worst side of the average. On the other hand, if entrepreneurs are told that only 1% of startups will succeed and still believe that they can beat the odds, they are more likely to succeed. The moment they believe in the statistics they would join the losers. All generals admit that to win a war, casualties are inevitable. Suppose again that they excel only at statistics (such as curves showing territories conquered vs. casualties), the moment that the generals wish to sacrifice a division for gains, they would be disappointed.

As reported by *Forbes*, a finance professor discovered[3] that the number of gold medals won at the Olympic Games is correlated with a nation's GDP, and therefore it is possible to predict the award of medals. While economic force lifts all boats, other factors are more sport-specific.

More money can do more things. If one cares to plot curves, a nation's consumption of beer, fuel, and toilet paper are all correlated with its GDP; any choice is arbitrary, and it is not always scientific. One may plot past investments vs. trophies won, but a football coach's challenge is in how to achieve better results with a given budget.[4] That is why, for many sports, tacticians are more valued than statisticians.

In almost every discipline the real experts are the equivalent of tacticians—except economics, which is still dominated by the equivalent of statisticians. Tacticians try to push many non-quantifiable constraints, and statisticians consider budget costs only. The former beat the curves, while the latter worship them.

For the sake of argument there must be a positive correlation between man's longevity vs. the world pesticide production, as both are the consequences of the expanding economy. Refined mathematic tools could easily prove the otherwise absurd proposition of more pesticides making man live longer!.

Another example is the diversification proposition (Section 4.1). One might say that more wealth leads to product diversification; but by the same logic one might also say that wealthy people live longer, and money must be the cure! While this is plausible, we know that to understand longevity, lifestyles and medicine play a direct role. For example, if a billionaire does not accept, for personal reasons, modern medical treatments, his wealth has no bearing on his longevity.

Curves cannot be distinguished apart if a correlation is "because of" or "in spite of." Even when it is clearly "because of," we still need to know the principal cause. For example, when a general wins a war it could be said that this is because of the food or water supplied to the soldiers, but although these are indeed indispensable they are clearly secondary causes.

Money should be more relevant for concerns such as poverty and famine, but even here the solution is not to simply throw cash to the poor. Atul Gawande[5] has reported the work by Jerry Sternin—a professor of nutrition at Tufts University—on how to cut malnutrition in poor countries by leveraging local skills that people can share.

Mainstream advice is well taken: there is no free lunch, but people in all walks of life constantly search for a cheaper and better lunch. For example, TripAdvisor enables its users more often stay on the better side of the quality–price curve than non-users.

This book aims to show that shrewd consumers, entrepreneurs, and investors—with the help of enabling institutions—can be on the better side of the average, and in the long run the curves also will improve. Collective curve-beating accounts for progress in technology and economic growth.

Recent work reported in *Nature*[6] also highlights the importance of studying dots instead of curves. Traditionally one would expect a country's competitiveness to be correlated with its GDP. Luciano Pietronero and his team found that the countries of the world scattered around a rising curve. Using their complex-network models they computed the driving force behind each dot (representing a country) and determined how it would evolve over time. By studying the dots and the deviations they could make predictions about the economies of the world. This work is in accord with our thesis that the dots are not random, and their deviations from correlation curves are the most interesting aspect on which to focus. We expect that many other examples—such as price/quality, reward/effort, and stock prices/underlying fundamentals—could be analyzed by following Pietronero's methodology.

Inferring from statistics for causality is a habit that goes beyond economics. For example, statistics professor Bjorn Lomborg has caused a sensation with his book *The Skeptic Environmentalist*, which denigrates the claim that pollution damages the environment. He plotted hundreds of curves to show that the environment improved with time,[7] despite the much-maligned pollution.

If the environment has not yet degraded too much, it is not due to Lomborg and the polluting industry's lobbyists who cheer him; it is due to the people who doubted the optimistic view and fought the polluters, which Lomborg's statistics cannot represent. If he and his supporters can convince a large fraction of the public, then the curve-plotting habit is not only useless but downright harmful. Even though some of his predictions are correct, we draw the exact opposite conclusion: these are not due to Lomborg's advocacy, but despite it.

Curve-plotting is a powerful tool for "hard" sciences where human cognition is not involved. If medical doctors plot data derived from smoking vs. cancer or fast food vs. obesity, the curves may reveal underlying causes. But bodily vs. mental activities must be dealt with by different methodologies. There is no reflexivity feedback loop in the smoking–cancer relation. The examples of quality vs. price, stock performance vs. risks, and environment vs. time reflexivity can impact the observed curves. Georges Soros—the most prominent proponent of the reflexivity principle—has pointed out that cognitive feedback loops impact financial markets and social sciences where the subject involves human cognition.

A theory, that more people believe in it would make it less true, cannot be good science. It might be said that this is unfair and cruel to the aspiration of economics becoming a hard science! What is meaningful cannot be studied as with hard sciences, and what can be studied (plotting curves) may be irrelevant, misleading, and even harmful, as it might degrade the very curves that theorists wish to celebrate. Even what is factually correct may still be bad science without adequate warning of potential abuses.

If many people find mainstream economics unsatisfactory, it is not that economists have done a bad job. With respect to other sciences, theirs is a much more difficult discipline with all types of reflexive feedback loops. Unlike atoms, economic agents' actions and scholars' theories might influence each other—and indeed, economists and their students are part of a complex belief system, and no-one can pretend to be a neutral observer.

Dots are obviously harder to investigate than curves, as each must be analyzed, and this forestalls sweeping statements. Practitioners know dots and scholars plot curves, and yet scholars can achieve much more with higher institutional designs that can have a big impact on the dots.

There are two fundamentally different methodologies: Newtonian and Darwinian. For the former, as in physics, dots are random and curves are essential; for the latter, the variations are more important, and the central types are only statistical shadows.

In biology, species in successive generations are practically identical; and yet the creationist theory can still be wrong on large-time scales. Natural Selection preferentially chooses fitter variants to pass on their genes, and complex life gradually emerged. Over time, the selected dots account for everything, while the center types followed.

Darwinian methodology is suitable whenever selection is important. Mainstream economics as it stands now is about averages and statistical curves. In a radical departure from the mainstream, our new theory is an enquiry about systematic bias of the dots from the curve. Hesitant variations are picked up by informational selection, and their cumulated effects are such that they constitute the entire economy on historical scales. Fundamental asymmetry also leads to the distinct attitude towards selection in consumer markets. For example, with the big data and AI, the introduction of PA (Chapter 3) discourages businesses from selecting weaker consumers from the average, and greatly enables selection power by consumers on businesses.

Mainstream economics cherishes averages, and the new theory advocates beating them. The former favors studying the curves, and the latter focuses on the dots and the cognitive gray-zone behind. Throughout history it is by beating averages and by consolidating advances that the economy has grown.

10.2 Allocation and creation

The allocation paradigm[8] states that "economics is concerned with the efficient allocation of scarce resources," which still underlies mainstream economics today. But it explains neither how resources come about nor how the economy evolves.

Berkeley economist Bradford Delong,[9] using the example of flour, reported that productivity has increased a hundredfold since 1500—but most items in the modern economy find no counterpart around 1500. New things emerge not as mere surprises, and precursors were known to pioneers.

While the economy grows, its composition also changes; the emergence of novelties is not simply due to allocating and reshuffling existent things. Resources do not rain down as mere externalities upon the economy; they are created indirectly via the allocation processes.

We propose the paradigm of allocation–creation, and this book aims to corroborate it. Allocation and creation are intimately related, as better exploitation (allocative) will stimulate further exploration (creative), and vice versa.

Physical objects—for example, stones scattered on the ground—do not grow, regardless of how we reallocate them. If an economy were to be filled with only existent objects and nothing new created, sooner or later it would settle into equilibrium—the best allocation. Throughout this book we have shown that resources such as magic pies, investments, and information content are different from physical objects; they grow or degrade, depending on how they are allocated.

Informational selection is apparently allocative, as to allocate we must gather information and determine/evaluate/compare alternatives. For example, consumers allocate their dollars among businesses, and better businesses may acquire larger shares of the market. Thus there is net resource creation when the market carries out the allocative job well.

Throughout this book we show that selection impacts on the agents via two pathways. The first pathway is due to reflexive feedback: upon market selection, businesses would dig into their hidden potential to improve their products.[10] The second pathway is competition, which reduces or eliminates mediocre businesses despite their product revision. The first pathway is voluntary and the second is involuntary, while the first pathway is creative and the second destructive. Among other examples, we have illustrated this point with discussion of B1 vs. B2, or post-production vs. prior-production (Section 3.5). B2 would represent creation, and B1 allocation.

The central hypothesis behind the new paradigm is that allocation and creation of resources cannot be separated. As a byproduct of incessant improvement on allocation, resources are inadvertently created.

Improving allocation is a process, and better selection tends to help improve allocation. The improvement on selection inevitably also reveals opportunities for resource creation. In other words, to better carry out their allocative tasks, agents must know more, and knowing more not only helps their intended allocative tasks, but also inadvertently allows them new opportunities. In the discussion of B1 and B2 (Section 3.5) we have emphasized that we cannot wait passively until businesses recognize the selection pressure and revise their products, but external help is needed to guide them in the B2 stage.

Because of this, allocation can never reach perfection; on new pies, the allocation processes must start again. The economy is full of new and old pies. Allocation on the old tends to be more efficient than that on the new, for the old are better known than the new. Imperfection diminishes here but emerges there, and though current tasks may be better allocated and exploited, improvement action will herald new tasks on which allocation is less well done.

Because of the inseparability, economic problems face indeterminacy. Mainstream economics bypasses this difficulty by setting artificial separation between the known and the unknown. We have argued that information and knowledge are distributed inhomogeneously across the population, and to determine who knows what we must deal with how people are connected, and the prevailing technologies. This book shows that inquiry into the gray zone is the key to understanding our economy.

Resource creation is less quantifiable than allocation; which is mainly due to the following asymmetry between them. Allocative acts are easier to measure because they deal with existent things, while creative acts are rarely observed except by the directly concerned. It is easier to account for the stabilizing forces on existent things than on the destabilizing forces caused by things newly created.

Allocation converts created resources to welfare in the economy; without allocation, creation matters only to its originators. Improving allocative efficiency allows innovators to go to farther frontiers, and the resulting fresh opportunities will in turn attract competitors good at allocation. But neither can go far without the other catching up. The economy consolidates the territories behind the advancing fronts and establishes the new center, and it is a never-ending process.

Resource creation is necessarily the job of insiders in their respective expertise—but we cannot conclude that the majority should sit and wait. Most people are outsiders, but everyone can be an insider in the right context. Entrepreneurs and policymakers can operate on what is equivalent to the level of a car-designer (Section 7.6), and institutional designs are the most effective way of assisting wealth creation, albeit indirectly.

The allocation paradigm often leads to the fixed-constraint fallacy: there is only that much to be reshuffled around. The new paradigm naturally accommodates positive and negative sum games, and the special case of zero-sum games is an exception. Scholars and policymakers should focus on how to foster favorable conditions that might promote positive sum games in the economy, which more probably arise with improved informational selection. Although scholars can rarely witness wealth creation directly, there is a great deal they can do to facilitate it, and indirectly they can contribute greatly with institutional designs.

New paradigm changes also affect the view on efficiency. Perfect efficiency can only be contemplated with fixed constraints that allow academics to carry out optimization. Our paradigm posits that efficiency can only be relative. For example, if a market is rendered a little more efficient, the constraints for that market also shift during the improvement process, hence no perfection. The most interesting questions still remain, however. How fast do the constraints shift, and how can entrepreneurs and policymakers anticipate the shifts?

Many consider that ecommerce matchmakers such as Amazon, Alibaba, and so on, improve consumer markets significantly, but that many traditional shops and workers were eliminated and the founders enrich themselves in the process. It is a matter of hot dispute whether they eliminate or create jobs. In the light of the new paradigm, we see that retail jobs (allocation) may be reduced, but the enhanced efficiency opens a great many new opportunities (creation). These two platforms cause both gains and losses in jobs, but we can draw a further conclusion: the lost jobs are more visible than the created jobs. This asymmetry makes people, especially policymakers, err more often on one side, and this leads to a systemic bias that the gains on new production side (creation) are less appreciated, but that the losses on the distribution side (allocation) are exaggerated.

Markets often appear to be just teasingly short of perfect efficiency, and we might have the illusion that with one more step the ideal would be within reach. This might lead some people to believe that perfect efficiency would serve as a sufficiently good approximation of the real world. This is perhaps why the ubiquitous information asymmetry in consumer markets is not thought to be important, as consumers and businesses are more or less prospering in the developed world. We show that while observation of the efficient functioning of markets might be factually correct, much potential could still be achieved by new institutional and structural innovations. We have illustrated this point with the example of PA. We can, of course, live without PA and still be happy, but its advent would bring about an enormous upheaval about which no business in the consumer markets could remain indifferent.

Therefore, our new theory focuses on exploring imperfections. It may seem that we were pursuing trifling gains (such as beating averages), but there is a much grander scope, in that chasing these imperfections can often lead us to new frontiers. In fact, this entire book is dedicated to the gray zone of seeming imperfections. The innumerous inefficiencies in front of us often seem small and insignificant, but the accumulated effects of pursuing them can be spectacular, and it in fact constitutes the future economy.

The fixed-constraint fallacy misleads some people to mistake a river for a pond. But however they might resemble each other, the river is fundamentally different from the pond, and the tiny gradient is the crucial telltale that the flowing river is an open system for which only dynamics befits its description.

Even though the two paradigms may, to an extent, account for the same facts, they could not be more different in worldviews. The new paradigm focuses on dynamics and endless evolution, taking into consideration both stabilizing and destabilizing forces. An expanding economy full of upheavals is the best evidence that it is not in equilibrium and never will be so.

Here we summarize the contrast between the two paradigms:

Old	vs.	new
Scarce resources		resources potentially unlimited
Fix constraints and then maximize,		watch out constraints shift
Zero sum game, allocate a fixed pie,		nonzero sum game, magic pie
Ideal limit with imperfections		they lead to new frontiers
Throw the gray out,		the gray is most interesting
Statisticians cherishing averages,		tacticians beat averages
Final state, equilibrium,		endless processes, evolution

The coevolution of man and the economy is powered by unlimited wants and skills. Had man's wants and skills ever stopped expanding, neoclassical economics would have been valid from that day forward. Much of the criticism raised in this book is well known in the history of economics. Many attributed the neoclassical paradigm to the misguided imitation of physics.

Physics works precisely only under ideal settings: for example, a projectile moving in a vacuum. In the real world, the theory holds true approximately, but the precision can be indefinitely improved if need be. No economist is so naïve as to believe in a perfect economy, but many consider that the real-world economy approximates the ideal limit—a world devoid of all sorts of imperfections. In textbooks, the ideal model is first presented, and imperfections are added in more advanced chapters. However, imperfect information is a misnomer, as if when the defects were removed, the economy would settle to the ideal world. In the view of the new paradigm, the ideal limit does not even exist, let alone reach it.

In physics, friction can be reduced and even removed, but what appears as friction in economics (imperfections, transaction costs, and so on), reveals, upon scrutiny, unending complexities. The so-called friction-free world cannot exist, since this would require us to know all future surprises. We have exposed this fallacy (Section 9.1) by using the example of communication costs, by which their reduction leads to a strange world where established business relationships are upset, and previous unimagined business models emerge. In physics, friction is treated as a nuisance, and it impedes our understanding the true laws of Nature. In economics, the seeming friction cannot be eliminated; on the contrary, we should pay utmost attention to it, as it might herald the future economy.

The new paradigm places mainstream economics on its head: what are considered as secondary deviants (such as dots, friction) are the most interesting; and what are considered as the main (such as averages, statistics) are merely statistical shadows. In the economy there is a gray zone extending from the well-known to the unknown. Allocative actions will change the composition of the known and the unknown; the gray zone is the focus in the new paradigm, and a dynamic picture emerges for an endless evolution.

Different paradigms are not only of academic interest; they lead to different world views that might be of importance in policy and real-world decisions. Our new paradigm also implies that when we solve current problems we must be aware of new problems that are inadvertently created, reflecting that any process would shift the initial constraints.

This is similar in spirit to what Kevin Kelly has said: "Technology creates our needs faster than it satisfies them;"[11] Bjorn Lomborg, however, has instead concluded that "our history shows that we solve more problems than we create."[12] DDT was thought to be a magic solution, and the pesticide industry today still insists that it did not create environmental problems. Solving obvious problems might create less visible problems that statistics would not reveal.

For Kelly, the constraints shift while we solve the current problems, whereas Lomborg succumbs to the fixed-constraint fallacy. Lomborg chases statistical shadows, and Kelly has the intimate insight of the technology precursors; their diverging world-views impact their own actions as well as those of their readers.

10.3 Relation to alternative theories

Mainstream economics has also evolved. Throughout the past few decades, many new ideas have extended much of the neoclassical framework, and economists' recent research has routinely considered information asymmetry, incomplete markets, and limited rationality. Indeed, many of the so-called shortcomings of mainstream economics mentioned in this book were known previously, and sometimes widely known. This book, however, proposes something new: a dynamic framework of a continuously evolving economy that may always appear imperfect, with many opportunities not yet appropriated. Therefore, the new approach never needs any benchmark to compare, as the economy is always out of equilibrium. We have found that these innovative concepts fit well with such an evolutionary view, hence we call it the new paradigm.

Mainstream economics is not a single block; there are many attempts to adapt it to the modern economy. For example, Joseph Stiglitz has systematically examined imperfect information issues across mainstream economics. In the previous work, however, to suit mathematical formulation, information is often treated as another variable with a price tag, thus bypassing the root issue of cognitive fallibilities. Such an approach has a serious drawback: how to price difficulties that are unknown?

Mainstream economics also deals with changes, but these are often treated as externalities. The economy would resettle after each external kick, and it might have jumped from one equilibrium to another, but the transition process remains a mystery. Our theory instead considers the economy in perpetual transition, without any end. It draws inspiration from some well-known alternative theories—notably, of behavioral economics and bounded rationality.

Behavioral economics considers a man more complex than *Homo economicus*; it reveals many failings of man as compared to the perfectly rational and informed representative agent. What is neglected is that man's secondary motives can really power our economy; the key lies in the conversion by informational selection.

We instead show that man's secondary motives are not merely a hindrance but can also be a productive force. What started out without an economic motive, once selected, can have considerable economic consequences. Secondary motives can be selected to yield unlimited wants and useful skills, and they play the role of unsung heroes in our economy.

Bounded rationality might have a negative connotation: but again, man is shown to be prone to failures, hence scholars regard him as bounded. While man may be a mediocre allocator, he can, on occasion, be a creator. On the bigger picture, what is deemed rational may not be so, and what is considered less optimal in a narrow sense might lead to upside surprises. It is the upside potential that the behavioral-bounded community neglects.

It is ironic that narrow rationality often impedes resource creation in the long run. Behavioral economics posits that man fails to optimize in many contexts; here we emphasize that he occasionally can do better than *Homo economicus*, and that his less than total selfishness can render him even better off. Man may not be as calculating as *Homo economicus*, but he is capable of outperforming it by other means.

For example, some do not heed the self-interest doctrine, and work on open-source software; some collaborate with each other, failing to acquire full legal protection and exposing them to the risk of exploitation. It is the risk-takers that make many innovations and startups possible, and the daring deviants contribute to the economy much more than the penny counters who optimize at every step.

Businesses are much more calculating than individuals, but what may seem surprising is that they do not necessarily optimize their profitability either, though not by miscalculation. Behavioral man may fail by negligence, and firms often deliberately forego profit optimization and divert huge resources to new spaces for uncertain future gains. All businesses consider the current constraints which they face—such as existing consumer demands, competition environments, financing terms, and so on—to be movable, and strive to move them. Current profitability vs. staking for the future always compete for resources, and the balance of the two is at the core of the corporate decision making.

Imperfect information economics and transaction cost economics focus on issues similar to those in this book. The main difference with respect to our approach is that they put a price on information, explicitly or implicitly. Information and knowledge often cannot be bought; some understand information better than others, and knowledge transmission depends on how people are connected. Moreover, information cannot be taken at face value, as its sources are often not reliable and can be intentionally distorted.

Labeling information problems as costs will forestall detailed investigation. The methodology still befits mainstream economics, since it writes off personal differences, social connections, skills, and institutional rules. For this reason we shall avoid using terminology such as transactions costs, human capital, social capital, and so on.

Theories of signaling and screening[13] offer many insights into how agents devise clever strategies to cope with the limitations of given information. The implicit assumption is that information asymmetry is a fixed obstacle, and people design sophisticated ways of circumventing obstacles. In our new view that not all information asymmetry problems are hard or impossible to reduce, we first focus on investigating whether some are reducible or not, and then design ways of changing these limitations. The most spectacular result of this approach concerns how to improve consumers' limited information capabilities, and we adopt the generalized, ubiquitous information asymmetry that is not even recognized as an obstacle in mainstream economics. Instead of appealing to consumers to be more vigilant, we focus on institutional designs (such as matchmakers) that may effectively enable consumers to be better informed.

In Section 9.4 we mentioned the signaling strategies of sexual selection in biology. Animals are also handicapped with limited information, and they devise ingenious strategies to reduce its effects. The key difference between informational selection and sexual selection is that for the former the information constraints might often shift, and for the latter the constraints can be regarded as fixed during millions of years of evolution.

The key difference with respect to previous theories on imperfect information is that whereas they work under the given constraints and try to best adapt to them, we work to push those constraints wherever possible.

In this new light, signaling and related theories account for what agents do best when constraints are fixed, while the new theory seeks ways to move constraints. These two approaches can be complementary, since economic agents often must do both at the same time. If all constraints turn out to be difficult to move, then mainstream economics and many of its recent extensions would be appropriate. Studying constraint-movability can shed light on the resilience of the constraints, since not all constraints are equally easy to move. In fact, much of the entrepreneurs' skill lies in identifying which constraints are easier to move and which moves produce better gains.

Austrian economics differs from mainstream economics by its emphasis on entrepreneurship, dynamic discoveries, and so on. While it considers that individual actions are the deciding factors in aggregate market behavior, here we emphasize that the most effective way of improving market performance is through interventions by third parties—notably by institutions built by private entrepreneurs.

The physics-imitation urge to quantify everything is the root cause of why many alternatives to the mainstream fall back to the embrace of equilibrium economics time and again. To produce a radically new paradigm that can rival the old, one must be equipped with a suitable methodology that can tackle the cognitive gray zone in the economy.

Almost everything discussed in this book can be co-opted into main-stream economics piecewise; mainstream economics proves to be resilient to the previous challenges without revising its root assumptions.[14] Thomas Kuhn has shown how hard it is for a new paradigm to dethrone an old one, and his favorite example concerns astronomy. Discoveries of celestial bodies could be made plausibly compatible with Ptolemy's geocentric theory; in the early years of Copernicus' heliocentric theory there were no clear winners, as both could fit the observations to some extent.

We may face a similar case in the impending paradigmatic revolution. Both the old and new paradigms may agree on the common starting point—the economy being somewhat efficient, agents more or less rational, and so on. But from here we part ways: mainstream economics leans towards statistic averages and the new theory focuses on the fringes, hence they are two fundamentally different theories. The new theory offers a coherent structure that can naturally accommodate many disparate elements, including those which mainstream critics propose, and a new paradigm emerges. This new paradigm has been anticipated in one way or another by many people, and especially by the following, whose work had greatly influenced this author.

Cliffe Leslie—an economist in the nineteenth century—was a well-known critic of the mainstream economics of his time led by David Ricardo and John Stuart Mill. Leslie advocated the historic methodology used by Adam Smith, and his view of economic evolution is still refreshingly relevant in today's world. In 1879 he postulated that "movement of the economic world has been one from simplicity to complexity, from uniformity to diversity, from unbroken custom to change, and, therefore, from known to unknown."[15]

Chicago economist Frank Knight first emphasized that uncertainty should not be confused with probability. In a similar vein, Santa Fe economist[16] Brian Arthur stated in his essay "The end of certainty in economics" that the economy "is not a well-ordered, gigantic machine. It is organic. At all levels, it contains pockets of indeterminacy." Herbert Simon and Amitai Etzioni have also proposed a realistic view of man and contemplated a worldview more complex than mainstream economics allowed. Tibor Scitovsky first explicitly considered the dual role of allocation and creation. The cumulative causation theory developed by Gunnar Myrdal emphasizes that the economy never settles in equilibrium with one institution kicking off the onset of another, without end.

Mark Granovetter's interdisciplinary work on the social aspects of the economy has had considerable influnce on my own thinking. For example, his study on embedded relationships offers much insight. Eric Beinhocker, in *The Origin of Wealth*, hints at a paradigm of allocation vs. growth, and advocates Darwinian methodology, and identifies the key elements of economic evolution as *differentiate*, *select*, and *amplify*. Bertin Martens—an economist at the EU commission—has also envisaged a similar research program.[17] James Buchanan and Victor Vanberg, in their paper "The market as creative process,"[18] state that "history as an open-ended evolving process and of a future that is not predetermined, merely waiting to be revealed" is "continuously originated by the pattern and sequence of human choice." Saras Sarasvathy[19] and colleagues identified three views of entrepreneurship: allocation, discovery, and creation—the last of which she advocates in earnest. Business consultant John Hagel and coauthors proclaim that "the edge is becoming the core."

The new paradigm was also anticipated by the Dartmouth economist Meir Kohn in his essay "Value and exchange," in which he states that the old value paradigm is being gradually challenged and that a new exchange paradigm should emphasize processes rather than the final state. It is worth quoting him further:

> At any moment the potential of the economy is not completely realized. Unexploited opportunities for mutually advantageous exchange abound. Indeed, the "potential" of the economy is not defined; it depends on the initiative and ingenuity of individuals. Individuals engaging in trading, innovation, and institutional change generate the process of growth, not only discovering potential but also creating it.

George Soros's reflexivity theory provides a methodological basis for this book. One of the most notable features of this theory is in its exposing the fixed-constraint fallacy, that under allocative actions, previously perceived constraints and targets shift. Soros has emphasized that human cognition, no matter whether it is right or wrong, can make the reality change. Mainstream finance theory considers the so-called fundamentals as given constraints, and Soros points out that human perception and misperception about them can change them.

Elsewhere we show[20] that reflexivity plays a much wider role in the economy than was previously recognized. Informational selection induces reflexive impacts on businesses who in turn are obliged to revise their products. In other words, reflexive impacts not only cause fallibilities but can also power economic growth, if channeled to productive ends.

The new role by reflexivity marks our important departure from Soros's theory. In fact, we attribute market expansion and economic growth to reflexive impacts on businesses; a stronger selection pressure by consumers would direct these impacts more towards product substance improvement than toward appearance.

10.4 Recapitulation

This book can be regarded as one long argument showing how resources allocation and creation together power market economies.

In view of the new paradigm, allocative actions take time, during which the constraints shift; hence both allocation and creation must be considered simultaneously. The two compete and complement each other, and together they drive the economy in an endless evolution.

In the economy there are always stabilizing (allocative) and destabilizing (creative) forces, hence it rarely settles into any equilibrium. Mainstream economics focuses only on the former, and this is probably the main cause of its mismatch with the real-world economy. In this book we show that when both stabilizing and destabilizing forces are accounted for, an alternative theory emerges that will have a fighting chance of rivalling mainstream economics.

The most important consequence of the new paradigm is that information capabilities can play the magic role of converting non-economic factors into material wealth, and the conversion is the true driving force behind economic growth.

Here we summarize the evidence in support of the new paradigm, and debunk the fixed-constraint fallacy.

Each agent faces a plethora of constraints: for example, among others, a consumer has limited information and a budget, an entrepreneur has a limited number of clients, and there is a competition environment. The boundaries between the known and the unknown are a special class of constraint that we have discussed extensively. Many constraints are related to the limited information capabilities, which can be improved, and thus cognitive boundaries can be pushed. Established business models and institutional settings represent the constraints within which agents operate. Prevailing technologies and political backgrounds can also be considered as constraints.

Besides these obvious constraints there are also mental constraints that may impinge on our economy. Notable examples are the dominant economic theories and prevailing consensus. Mainstream finance even considers stocks' average yields as given conditions, and builds an entire optimal portfolio theory on the assumption that these are stable.

Throughout this book we have argued that most of these constraints can, to an extent, be moved. We focus on why and how they move, who causes the moving, and the consequences. Some moves are deliberately planned and painstakingly pursued, and some occur inadvertently.[21]

Businesses may even welcome constraints; for example, in the defensive region (Section 1.3) they might set up artificial barriers to downgrade consumers' information capabilities. Constraints moving can result in magic-pie growth or reduction, and even when it becomes bigger, some stakeholders may still lose, hence not everyone embraces constraint-moving.

As a first application of the new paradigm (Section 1.2) we have revised the mainstream supply–demand law by allowing its traditional fixed constraints—quality and information capabilities—to move. If these two constraints can shift, businesses have incentives to revise their products' quality, and similarly to the revision of prices. Enhanced selection pressure makes better allocation of consumers' budget, and at the same time it can make the magic pie bigger.

We have shown that stronger selection is the key to improving resources allocation and creation; hence much space is devoted to how the selection pressure can be enhanced (Section 2.2). Your cognitive constraints and mine are different, and information matchmakers can leverage the difference and enable consumers collectively to be more informed (Section 2.3).

As a response to the expansion of consumers' cognitive boundaries, businesses might reset their production constraints by revising their products and even business models, and we can speak of a general reflexive impact on the businesses that gives rise to product diversification (Section 4.1). There is a systematic trend of consumers being more and more informed and this trend, instead of taking us closer to perfect understanding, leads to product diversification.

Financial markets (Section 5.3) can contribute to resources creation in the real-world economy by selecting and financing investment opportunities. Better selection leads to better allocation of investment capital, allowing worthwhile projects to be funded; hence it contributes to wealth creation. Allocation and creation manifest in the financial markets via the symbiotic relationship between investors and businesses seeking capitals.

Mainstream finance theory considers a fixed set of investments and investors. These fixed constraints allow for computing the optimal allocation. To debunk the fixed-constraint fallacy in this context, we argue that if all investors heed the mainstream doctrine, selection work that no mathematics can describe would be neglected, and stocks without intense selective scrutiny would rapidly degrade in quality (and hence yields). What mainstream economics considers as given, our new theory regards as a precarious compromise that can either degrade or improve, depending on the prevailing selection capabilities of the investors.

Two paradigms are similarly confronted on the markets for information. Search is allocative: facing the vast repository of online information, each user should access the most relevant information required. Leading search engines too are often unwitting victims of the fixed-constraint fallacy, as they seem to consider existent information as given, and ignore the fact that search algorithms can have a major impact on information genesis (Section 6.3).

This is like optimization within a fixed box by mainstream economics. If they could realize that the current content is merely a tiny fraction of what the authors could provide, they would enable feedback from the searchers to the authors. If they decline to take part in moving constraints in productive ways, there are also inadvertent constraints shifts that downgrade the information ecology. Prevailing search algorithms can build up systematic biases (Section 5.3), making rare niches harder to find.

These three types of market share the common theme that the most important of the constraints in the economy are the attributes of man, which were regarded as given in the prevailing doctrines. We have shown that both wants and skills can expand without end, and deliberate selection on many of man's secondary motives can convert a fraction of them to power the economy.

In the mainstream view, human agents passively adapt to constraints imposed by the powers that be, and resign to their fate. Following the new paradigm, human agents must not only cope with current constraints, but also take an active part in moving constraints. This gives rise to additional leverage and uncertainties. Facing the possibility of driving constraints by deliberate effort, there are many ways to move constraints, and success or failure is often dependeant on the choices. Economic agents must tackle more complex tasks (a larger space to choose alternatives) than optimization under constraints.

The new paradigm also implies a novel working agenda for both practitioners and policymakers. They would test all seeming constraints, and then choose to move those that can bring benefits. Some constraints are easier to move, others are more difficult, some moves immediately bear fruit, and some require long-term foresight.

In short, the new paradigm posits that human agents can be influential in deciding their own destiny, and if we squander our opportunities we have only ourselves to blame. Regarding the future, this book offers an alternative view by which each agent, with his own expertise, can anticipate future opportunities. His current effort contributes to the realization of future events. If agents can anticipate only a little better than the average, the new view is dramatically different from the mainstream equilibrium view, as it offers a dynamic perspective that can account for the vibrant economy evolving without end. These differentiated ways of seeing the future make the future arrive a little sooner.

In mainstream economics views, human agents try randomly, and some succeed and some fail. Our new view is that each strives to do better, and while some succeed and some fail, their efforts, skills, and talent, as well as chance, each play a role. Outside observers should not sit on the sidelines, but may help the agents by rendering institutional settings more conductive to successes.

To summarize the key difference between the two paradigms, we reiterate that mainstream economics assumes a river to be a pond. But a river is more difficult to describe, as we must delve deeper into the visible body of water in order to probe its sources and outlets. The new paradigm abandons optimization, and begins to enquire into the real-world economy as a dynamic process.

Notes and References

Preface

1. The unfortunate uselessness of most "state-of-the-art" academic monetary economics: https://voxeu.org/article/macroeconomics-crisis-irrelevance.
2. Frank Hahn (1973), *On the Nature of Equilibrium in Economics*, Cambridge University Press, pp. 14–15.
3. https://en.wikipedia.org/wiki/Kardar–Parisi–Zhang_equation.
4. D. Challet, M. Marsili, and Y.-C. Zhang (2005), *Minority Games*, Oxford University Press.

Introduction

1. <http://rodrik.typepad.com/dani_rodriks_weblog/2008/02/one-economics-a.html>.
2. G. Soros: "The assumption of perfect knowledge became untenable and it was replaced by a methodological device which was invented by my professor at the London School of Economics, Lionel Robbins, who asserted that the task of economics is to study the relationship between supply and demand; therefore, it must take supply and demand as given. This methodological device has managed to protect equilibrium theory from the onslaught of reality down to the present day". <http://mertsahinoglu.com/research/the-theory-of-reflexivity-by-george-soros/>.

Chapter 1: Magic Pie

1. Standard economics textbooks write off information by portraying the supply–demand law (quantity vs. price) with simple examples such as oranges or apples, but rarely use more complex examples in which information challenges arise.
2. After simple examples (oranges, apples, or ice cream), mainstream textbooks, without hesitation, generalize their conclusions with more complex products. For example, Berkeley economist Bradford Delong derides Gregory Mankiw's attempt to compare healthcare with groceries: <http://www.bradford-delong.com/2009/06/the-public-plan-for-health-insurance-in-which-greg-mankiw-confesses-to-remarkable-ignorance-and-asks-a-question-that-we-answ.html>. Healthcare is particularly prone to information deficiency. For example, Winand Emons has cited a study of the Canton of Ticino in Switzerland in which about 33% of surgeries were unnecessary: *RAND Journal of Economics*, 28 (1997), 107. Over-treatments are far from being

limited to Switzerland, and further examples will be discussed in subsequent chapters.

3. George Akerlof, "The market for lemons: qualitative uncertainty and the market mechanism," *Quarterly Journal of Economics*, 84 (1970), 488.

4. Vendors and producers suffer information asymmetry of another kind: what consumers really want. See Sections 1.4, 2.5, and 2.6.

5. Limited infocap is similar to the bounded rationality and "cognitive limitations" proposed by Herbert A. Simon in *Models of Man*, but infocap is better suited for our purposes in this book. It is impossible to define infocap as precisely as in communication sciences, but real-life examples show significant differences in infocap both for different people on the same task and a given person on different tasks. A gray-scale represented by a continuous parameter is convenient for studying shifts, as different forces try to shift infocap in different directions.

6. Ubiquitous information deficiency is consistent with Soros's "radical fallibility" hypothesis: humans have only a patchy understanding of the reality in most circumstances. We extend Soros's work by proposing the relative fallibility concept that economic agents understand some and miss some: the glass is half empty or half full. By "half empty" we emphasize man's fallibilities, and by "half full," capabilities. Agents do not fully understand the market that they face, but they are rarely totally ignorant about it; the key lies in the relative degree of capabilities.

7. "How Design Thinking Transformed Airbnb from a Failing Startup to a Billion Dollar Business:" <http://firstround.com/review/How-design-thinking-transformed-Airbnb-from-failing-startup-to-billion-dollar-business/>.

8. In D. W. Carlton and J. M. Perloff, *Modern Industry Organizations* (1989), the authors discuss tourists vs. locals as examples of inhomogeneous infocap.

9. See, for example, the study concerning the price spread of ketchup in London: Eric Beinhocker, *The Origin of Wealth*, Random House (2007), p. 62.

10. In traditional supply–demand laws, prices and quantity are regarded as variables, and it is puzzling that information and quality were considered given and fixed for so long without relaxing the initial simplifying assumptions.

11. Y.-C. Zhang, "Supply and demand law under limited information," *Physica A: Statistical Mechanics and its Applications*, 350 (2005), 500–32.

12. On consumer markets it is often the vendor and the producer who take the initiative in deciding price and quality, but the sale volume depends on the buyers' acceptance.

13. This is similar to the Cournot Competition, but with imperfect information: see H. Liao, et al., "Firm competition in a probabilistic framework of consumer choice," *Physica A: Statistical Mechanics and its Applications*, 400 (2014), 47–56.

14. See, for example, the standard textbook by Hal Varian, *Intermediate Micro-Economics*, Norton (1991).

15. T. Scitovsky was the first to point out that limited infocap will lead to consumers finding it difficult to distinguish competing offers: "Benefits of asymmetric markets," *Journal of Economic Perspectives*, 14 (1990), 135–48.

16. This is a recurrent dilemma throughout this book: see embedded relationships in Section 6.3 for an analogy with investing across many stocks vs. concentrating on a few. See A. Capocci and Y.-C. Zhang, "Driving force in investment," *International Journal of Theoretical and Applied Finance*, 3 (2000), 511; and for a classic paper, George J. Stigler, "The economics of information" *Journal of Political Economy*, 69 (1961), 213. It can be related to what Cambridge economist Joan Robinson has termed "imperfect competition."

17. H. Liao et al., "Firm competition in a probabilistic framework of consumer choice."

18. In business-school literature it is called Product Life Cycle; see, for example, Theodore Levitt, "Exploit the Product Life Cycle," *Harvard Business Review*, 43 (1965), 81–94. Note the important difference that in PLC literature the horizontal axis is time, whereas in our Figure it is infocap. However, since infocap normally arises with time, though not linearly, the two concepts are closely related. While PLC is an important concept in business schools, it is not grounded in the mainstream supply–demand law, and seems to be an ad hoc observation. On the contrary, our Figure is a natural consequence of the new supply–demand relation.

19. A report shows that during a period of five years there is a broad trend showing that consumers are better informed due to the Internet: <http://www.pewinternet.org/2014/12/08/better-informed/pi_2014-12-08_better-informed-06/>.

20. Charlie Barone, "The best customers are informed customers," *Automotive Body Repair News*, 39 (2000), 26.

21. "Car dealers might have to deal with informed customers! That must be illegal!": <http://www.techdirt.com>, February 13, 2012.

22. For a representative work, see Gary Becker and George J. Stigler, "De gustibus non est disputandum," *The American Economic Review*, 67 (1977), 76–90.

23. Al Ries and Laura Ries, *The 22 Immutable Laws of Branding*, Collins (2002).

24. The book edited by Kyle Bagwell summarizes the state of the art: *The Economics of Advertising*, Edward Elgar (2001).

25. The *Businessweek* report "The great rebate runaround," by Brian Grow, December 5, 2005, analyzes many aspects of rebate marketing strategy. For example, it was found that about 40% of all rebates were never redeemed, and some fulfilment firms were not shy to boast that successful redemption rates can be as low as 10%.

26. Tibor Scitovsky, *Joyless Economy: The Psychology of Human Satisfaction*, Oxford University Press (1977).

27. The US–Russian economist Evsey Domar was the first to propose this asymmetry: *The Blind Men and the Elephant: An Essay on Isms, in Capitalism, Socialism, and Serfdom*, Cambridge University Press (1989), pp. 29–46.

28. L. Lü, M. Medo, Y.-C. Zhang, and D. Challet, "Emergence of product differentiation from consumer heterogeneity and asymmetric information," *The European Physical Journal*, 64 (2008), 293–300. M. Medo and Y.-C. Zhang, "Firm competition in a probabilistic framework of consumer choice," *Physica A: Statistical Mechanics and its Applications*, 387 (2008), 2889.

29. From the cognitive point of view the magic pie represents our explicit wants and the new pie implicit wants. Explicit wants are those we know we need but cannot be sure of quality, and implicit wants are those we may not know yet. See the discussions in Sections 2.5, 2.6, and 7.2.

30. Cory Doctorow, "Giving It Away", Forbes: <https://www.forbes.com/2006/11/30/cory-doctorow-copyright-tech-media_cz_cd_books06_1201doctorow.html#480b399078c2>, November 30, 2006.

31. Infocap on new pies is even weaker. It is convenient to separate transactions into two categories: the first consists of whatever you consciously plan to buy; the second includes advertisement-induced purchases. You probably have higher infocap over items in the first category than those in the second. The reason we succumb to advertisements and buy something as a result is that they occasionally fill our cognitive gaps. The chance is small for an arbitrary advertisement to catch our fancy.

Chapter 2: Matchmakers

1. There are two well-known business models for matchmakers: merchant and agency. The merchant model is not much different from the middleman (discussed in Section 2.1) who holds a block of offers, just as Expedia sometimes holds many hotel rooms. The agency model moves the matchmaker closer to consumers, as it is no longer eager to push them to buy a specific offer among its stock, but allows them to choose. In the online travel industry there is a trend for matchmakers to transit from the merchant model to the agency model; that is, to be closer to consumers.

2. D. Spulber, *Market Microstructure: Intermediaries and the Theory of the Firm*, Cambridge University Press (2004). Spulber points out that middlemen emerge when they improve efficiency. In the Preface he states that "firms are formed when the gains from intermediated exchange exceed the gains from direct exchange".

3. The relationships can be four-way or even more. For example, eBay.com and its buyers and sellers are in a three-way relationship; and Buysafe.com can check and certify the authenticity of products for consumers.

Therefore, within the relatively simple three-way structure there can be embedded other, more complex structures.

4. The sudden appearance of many different business models might be compared with the Cambrian Explosion—the sudden appearance of numerous animal species in a geologically short period between 542 and 520 million years ago. See Ray Kurzweil, *The Singularity is Near: When Humans Transcend Biology*, Penguin (2006).

5. Another role by matchmakers is indirectly related to infocap improvement. If you are not a frequent patron, a business has no incentive to treat you particularly well. But a matchmaker acts effectively as a large union for its members, and a business must treat all members of the "union" as if they were repeat patrons. Thus, besides infocap improvement, the mere fact of binding consumers together creates value for consumers.

6. Platform matchmakers differ from expert matchmakers, as the former's value is proportional to N squared and the latter's is only linearly in N. See Section 2.2 for a discussion of why platform matchmakers tend to be ultra-monopolies.

7. For examples, Curetogether.com and patientlikeme.com are such experience-sharing websites.

8. *Businessweek* interview with Hal Varian, May 13, 2001: "Internet auction site eBay Inc.'s business was not actually as novel as some people thought. It is really just electronic classifieds, he says. What made the service more valuable than the Sunday paper was that with many more buyers and sellers available, pricing became incredibly efficient, and the thing snowballed. His lesson: Look for something that has worked in the past and figure out a way to supercharge it using the Net." <http://www.bloomberg.com/bw/stories/2001-05-13/hal-varian>.

9. Paul Resnick et al., "Herding the mob." <http://wired.com>, March 2007.

10. M. Luca, Harvard Business School working paper, 2011, "Reviews, reputation, and revenue: The case of Yelp.com."

11. See the report on Techcrunch.com by G. Ferenstein, "Berkeley study: Half-star change in Yelp rating can make or break a restaurant," September 2, 2012. "Two Berkeley economists have found that the tiniest changes in online restaurant reviews can make or break a restaurant. A simple half-star improvement on Yelp's 5-star rating makes it 30–49% more likely that a restaurant will sell out its evening seats. Online reviews, the researchers conclude, play an increasingly important role in how consumers judge the quality of goods and services." <http://techcrunch.com/author/gregory-ferenstein/>.

12. See the review sites: <http://www.ratemyteachers.com>; <http://www.ratemyprofessors.com>.

13. Questionabledoctors.org was ordered to shut down after pressure from the powerful healthcare lobby; see its tuned-down version: <http://www.citizen.

org/physicianaccountability>. The lobby complained that sometimes patients' critique is questionable. Humans can err, but doctors can sometimes make very serious mistakes, and patients may err in their reviews. See "Doctors try to silence negative reviews from patients" on the Ars Technica website: <http://arstechnica.com/business/2009/03/doctors-try-to-silence-negative-reviews-from-patients/>.

14. Surgeon and author A. Gawande's book *Complications: A Surgeon's Notes on an Imperfect Science* (Picador Publishing, 2003) tells an insider's stories.

15. For example, the watchdog site factcheck.org tracks accountability of politicians and businesses.

16. *San Francisco Chronicle*, September 3, 2006: "Amateur reviews changing approach of small businesses online ratings: It started with restaurants, and now all manner of enterprises find themselves subject to customer opinions."

17. This is known as Metcalfe's law, which states that a network's value is proportional to the square of the number of the nodes. Platform matchmakers' values are N squared, but for expert matchmakers such as CNET and ZAGAT their value is linear in N.

18. In real life there are still many competing search engines, but they are usually not standalone businesses and have strong backers with other lucrative activities (such as Microsoft's Bing.com). Competitors often target slightly differentiated groups, hence they are de facto monopolies in their own niches.

19. Consumers are less calculating than businesses, but have many more motivations and propensities than firms have (Section 7.1).

20. As per fundamental asymmetry, any given item of a consumer's wants is easily saturated, and they become less exploitable by more than one business. Business offers, on the other hand, are much easier to scale up. In Chapter 1 we remarked that McDonald's can double their burger production more easily than its customers can double their appetite.

21. Recommendation precision is also related to infocap; with the distinction that now consumers are being passively served. Netflix commands a higher precision than Google's advertisements, whose task is much more difficult: queries and advertisements can be of any genre, and matching is more difficult than the Netflix challenge where matching is within a relatively small pool (movies and viewers).

22. See <www.mediapost.com/publications/article/258605/how-google-will-know-you-better-than-you-know-your.html>.

23. It is easier to charge for new pies than for existing ones. The reason is twofold. First, new pies are created entirely due to matchmakers' initiatives, hence it is easier for matchmakers to lay claim. Second, consumers' infocap on new pies tends to be still weaker than for existing pies, as discussed in

Chapter 1. Businesses enjoy bigger margins on new pies and are hence more willing to share the profits with matchmakers.

24. Mediating between specialists and generalists poses a dilemma for match-makers. In principle, whoever benefits from the service should pay, but collecting payments from many consumers is a difficult task. This difficulty is related to Ronald Coase's theory of transaction costs.

25. For example, Pandora.com, an online radio site, tried the subscription model when it started in 2005. "That lasted all of three weeks," said Tim Westergren, the founder. "It was pretty clear there was no future in that and the only real option was to go free."

26. This puzzle is one of many indirect consequences of fundamental asymmetry.

27. See *Businessweek*, "Downloads: The next generation, big music worst mistake," February 16, 2004.

28. C. C. Miller, "Ad revenue on the Web? No sure bet," *New York Times*, May 25, 2009.

29. "This tech bubble is different," *Businessweek*, April 14, 2011.

30. See Tom Staffold, *Mind Hacks: Tips & Tools for Using Your Brain*, O'Reilly Media (2004).

31. The credit card industry is dominated by a few global giants such as MasterCard, Visa, and American Express. Their data dwarf those of Amazon and other recommendation pioneers. How often do you shop on Amazon, and how often do you buy anything at all using credit cards? Current card issuers now comfortably collect fees and will not be bothered with the complex issue of figuring out clients' preferences based on purchase patterns—or so they may think. Such a potential awaits innovators who would be paid entirely by new pies which they could enable. Current card companies sit on a huge untapped profit potential that is magnitudes larger than the fees they charge now. Besides transiting from a service vendor to an enticement matchmaker, the industry can benefit greatly from the coming trend in mobile payments. On the surface, mobile payments seem to be a convenient extension of credit cards, but they allows businesses to acquire much more location and time-sensitive data. The business model combining the enticement matchmaker and the mobile payment technology will have far-reaching impacts on consumer markets.

Chapter 3: Personal Assistant

1. See <http://techcrunch.com/2014/12/13/what-artificial-intelligence-is-not/>.

2. See, for example, <http://wired.com/2014/05/anonymous-apps/>. On smart-phones, many apps serve as small independent gateways to a service, but PA would eventually reduce the large number of access points to only one by calling on necessary apps where appropriate.

3. By 2016, Apple Store had more than a million apps, reminiscent of the early days of the Internet with millions of websites, but a coherent picture is lacking.

4. Push and pull are the well-known models for supply chain management. See, for example, John Hagel III, John Seely Brown, and Lang Davison, *The Power of Pull*, Basic Books (2012). This is similar in spirit, but in our context we focus on how consumers and businesses find each other. IT consultant John Hagel and colleagues observed that information content is also shifting from push to pull.

5. Businesses and consumers rank products with different preferences: what a business likes the most may not be the best for a given consumer, and vice versa. Their relationship is partially overlapping and partially conflicting.

6. Similar stories arise in China, such as Pinduoduo and Netease's Yanxuan. The former does well for p only, whereas the latter for both p and Q.

7. Here we consider privacy concerns for legitimate businesses only. For illicit offenders who engage in spams, phishing, and identity theft with criminal intentions, the privacy concern is much more serious and is beyond the scope of this book.

8. That why we discussed average infocap and average product quality in Chapter 1.

9. The issue of connected PAs of all consumers is an extensive subject that will be discussed in a separate book dealing with all scenarios when PAs are connected. Among other factors, consumers connected via their PAs would be able to evaluate all products, much more than would IDOL (Section 2.3).

10. There are still earlier stages beyond B2; for example, how business plans are drawn, innovation fostered, companies incorporated, talents discovered, personnel trained, and so on.

11. Seth's Blog, "Ad Blocker": <http://sethgodin.typepad.com/seths_blog/2015/09/ad-blocking.html>.

12. Kevin Kelly, *New Rules for the New Economy: 10 Radical Strategies for a Connected World*, Viking Adult (1998), chapter 10.

Chapter 4: Diversification

1. Barry Schwarz, *The Paradox of Choice: Why More Is Less*, Harper Perennial (2005).

2. Alfred Marshall, *Principles of Economics*, Cosimo Classics, 8th edn. (2009), p. 355.

3. E. Von Hippel, *Democratizing Innovation*, MIT Press (2006), chapter 1,

4. Chris Anderson, *The Long Tail: How Endless Choice is Creating Unlimited Demand*, Cornerstone Digital (2010).

5. Erik Brynjolfsson, Yu Jeffrey Hu, and Michael D. Smith, "The longer tail: The changing shape of Amazon's sales distribution curve," *Social Science Research Network*, September 2010: <http://ssrn.com/abstract=1679991>.

6. Some people have questioned whether IT is always good for the long tail. For example, Tom Woodward has observed: "Online merchants such as Amazon, iTunes, and Netflix may stock more items than your local book, CD, or video store, but they are no friend to 'niche culture':" Bionic Teaching blog: <https://bionicteaching.com/>, April 6, 2009. Tom Slee has also voiced doubt in "Online monoculture and the end of the niche:" <http://tomslee.net/blog/page/28>. See also M. Kirkpatrick, "Are recommendation engines a threat to the long tail?"

7. In the extreme case when systemic diversity (person-to-person variation) approaches zero, recommender systems stop work.

8. Toffler's *The Third Wave* describes the transition in developed countries from Industrial Age society, which he calls the "second wave," to the Information Age, the "third wave."

9. John Kenneth Galbraith, *The Affluent Society*, Houghton Mifflin (1958).

10. Jack Trout, *Differentiate or Die: Survival in Our Era of Killer Competition*, Wiley (2001).

11. Brynjolfsson et al. "The longer tail."

12. B. Tedeschi, "Small merchants gain large presence on Web," *New York Times*, December 3, 2007.

13. See <http://www.antipope.org/charlie/blog-static/2011/12/the-coming-retail-apocalypse-s.html>.

14. Michael Pollan, *The Omnivore's Dilemma*, Penguin (2007). See also <https://cowpool.org/whycowpool>.

15. Charles Fishman, *The Wal-Mart Effect: How the World's Most Powerful Company Really Works*, Penguin (2006).

16. Dan Koeppel, *Banana: The Fate of the Fruit That Changed the World*, Plume (2008).

17. Megan McArdle, "What's wrong with chain restaurants?" *The Atlantic*, May 6, 2008.

18. Michael Luca, *Reviews, Reputation, and Revenue: The Case of Yelp.com*: <http://www.yelpblog.com/2011/10/harvard-study-yelp-drives-demand-for-independent-restaurants>.

19. Maxwell Wessel, "Why big companies can't innovate," *Harvard Business Review*, September 27, 2012.

20. "Editorial," *The Economist*, January 24, 2007.

21. See "The Chinese are spending 43% less time watching TV and 45% more time surfing the Net," <http://thenextweb.com/asia/2011/05/30>.

22. See also <http://techcrunch.com/2011/07/03/the-power-of-pull>.

23. Clayton M. Christensen, *Innovator's Dilemma: When New Technologies Cause Great Firms to Fail*, HBR Press (2013).

24. Marketing Guru Philip Kotler has said: "Good companies will meet needs; great companies will create markets;" cited in *The 75 Greatest Management Decisions Ever Made*, MJF Books (2002), p. 37.

25. Amit Kapur, "The future will be personalized," *Techcrunch*, November 16, 2010.
26. Emile Durkheim, in his well-known work *The Division Of Labor In Society*, advocated the role of the division of labor in the progress of a society.
27. Craig Canine, "Building a better banana," *Smithsonian Magazine*, October 2005.
28. Emma Young, "Featherless chicken creates a flap," *New Scientist*, May 21, 2002.
29. See Kevin Kelly, "Increasing Diversity:" <http://kk.org/thetechnium/increasing-dive>.
30. See <http://www.seedsavers.org>. See also CBS News, "'Doomsday' seed vault opens in Norway": <https://www.cbsnews.com/news/doomsday-seed-vault-opens-in-norway/>, February 25, 2008.
31. Donald L. Barlett and James B. Steele, "Monsanto's harvest of fear," *Vanity Fair*, May 2008.
32. Thomas H. Maugh II, "Researchers say a language disappears every two weeks," *Los Angeles Times*, September 19, 2007.
33. Cass R. Sunstein, "Paradoxes of the regulatory state," *University of Chicago Law Review*, 157 (1990), 410.
34. See statistician Bjorn Lomborg's *The Skeptical Environmentalist*. Cambridge University Press (2001).
35. David Hillis has said that less than 10% of all species on Earth have ever been described: BBC News, "Internet helps write the book of life," January 9, 2003.
36. Samuel Loewenberg, "Precaution is for Europeans," *New York Times*, May 18, 2003.

Chapter 5: Financial Markets

1. Paul Krugman, "There'll always be a Soros," *Fortune Magazine*, March 30, 1998.
2. George Soros, *Alchemy of Finance*, Simon & Schuster (1987).
3. Soros, *Alchemy*, p. 54.
4. John M. Keynes, *The General Theory of Employment, Interest and Money*, Macmillan. (1936), chapter 12.
5. Paul A. Samuelson "It is not easy to get rich in Las Vegas, at Churchill Downs or at the local Merrill Lynch office."
6. Keynes, *The General Theory of Employment*, pp.161–2.
7. D. Levy on Tom Cover's work: "Universal portfolios take investors back to the future," *Stanford Report*, April 2000.
8. At the same infocap level per task, more diversification is more desirable; or at the same diversification level more depth allows you to better investigate the underlying opportunity. Infocap being limited, we must make a compromise.
9. Andrea Capocci and Y.-C. Zhang, "Driving force in investment," *International Journal of Theoretical and Applied Finance*, 3 (2000), 511.

10. In Thomas Friedman's *The Lexus and The Olive Tree* (p. 118) he cites a story about how Goldman Sachs analysts covered typically seventy-five stocks in 1967, but three decades after the coverage it was reduced to twelve stocks on average.

11. J. Micklethwait and A. Woodridge, *Right Nation,* Penguin (2005). They reported that 19% of Americans believe that they are among the richest 1%. See also D. Brooks, "Superiority Complex," *The Atlantic*, November 2002.

12. M. Medo, Y. M. Pis'mak, and Y.-C. Zhang, "Diversification and limited information in the Kelly game," *Physica A: Statistical Mechanics and its Applications*, 387 (2008), 6151.

13. Think of a chef who never eats his own dishes; the well-known agency problem will produce terrible food.

14. "Ironically, it is the very use of the crash-free Black-Scholes model that destabilized the market:" J.-P. Bouchaud, "Economics needs a scientific revolution," *Nature*, 455 (2008), 1181.

15. M. Medo, C. H. Yeung, and Y.-C. Zhang, "How to quantify the influence of correlations on investment diversification," *International Review of Financial Analysis*, 18 (2009), 34.

16. There are various versions of financial market efficiency in the literature. Here we define efficiency as a measure of exploitable profit opportunities. A market is said to be more efficient than another if it has fewer remaining opportunities. It is a relative concept, as some investors cannot or will not detect exploitable opportunities, and for them the market is efficient. We posit that efficiency aims to promote symbiosis. Whether or not a market is made more efficient, the sole criterion will be whether or not symbiosis is improved. Absolute efficiency does not exist, as our economy can evolve without end.

17. Y.-C. Zhang, "Toward a theory of marginally efficient markets," *Physica A: Statistical Mechanics and its Applications*, 269 (1999), 30.

18. J. D. Farmer, "Toward agent-based models for investment," *AIMR Conference Proceedings*, ed. R. Max Darnell (2001), p. 64.

19. The most used device in mainstream economics, ceteris paribus, writes off the interplay and reflexivity.

20. Rajnish Mehra and Edward C. Prescott, "The equity premium: A puzzle," *Journal of Monetary Economics*, 15(1985), 145.

21. W. Buffett said further: "Ships will sail around the world but the Flat Earth Society will flourish. There will continue to be wide discrepancies between price and value in the marketplace, and those who read their Graham & Dodd will continue to prosper." <http://www.businessinsider.com/warren-buffett-on-efficient-market-hypothesis-2010-12>.

22. "In Defence of the Dismal Science", *The Economist*, August 2009, quotes R. Lucas as saying: "The main lesson we should take away from the EMH for policymaking purposes is the futility of trying to deal with crises and

recessions by finding central bankers and regulators who can identify and puncture bubbles."

23. See "Andreas Treichl of Erste Bank thinks banks should be kept small and simple," *The Economist*, September 3, 2009.

24. E. L. Andrews, "Greenspan concedes error on regulation," *New York Times*, October 23, 2008: "Those of us who have looked to the self-interest of lending institutions to protect shareholders' equity, myself included, are in a state of shocked disbelief."

Chapter 6: Information Markets

1. See "Top Google result gets 36.4% of clicks." http://searchenginewatch.com/sew/news/2049695/top-google-result-gets-364-clicks-study

2. D. Kirkpatrick, "Book agent's buying fuels concern on influencing best-seller lists", *New York Times*, August 23, 2000.

3. A.-L. Barabási and R. Albert, "Emergence of scaling in random networks," *Science*, 286 (1999), 509.

4. D. Dranove, D. Kessler, M. McClellan, and M. Satterthwaite, "Is more information better? The effects of health care quality report cards," *Journal of Political Economy*, 111 (2003), 555.

5. M. Ceaser, "Colombia military accused of killing civilians," *SFGate*, February 22, 2009. Human rights organizations stated that more than 1,000 civilians have been killed by soldiers and police in recent years to bolster rebel casualties.

6. Jared M. Diamond, *Guns, Germs, and Steel: The Fate of Human Societies*, Norton (1999).

7. John Ellis, "Will Google Instant kill the long tail?" *Search Engine Land*, September 8, 2010. Ryan Singel, "What's in a search, if you don't hit the search button?" <http://wired.com>, September 2010.

8. "Google's robotic recipe search favors SEO over good food:" <http:techcrunch.com/2011/03/26/googles-robotic-recipe-search-favors-seo-over-good-food/>.

9. T. Zhou, Z. Kuscsik, J. Liu, M. Medo, J. Wakeling, and Y.-C. Zhang, "Solving the apparent diversity–accuracy dilemma of recommender systems," *Proceedings of the National Academy of Sciences*, 107 (2010), 4511.

Chapter 7: From Markets to the Economy

1. Mark Granovetter and Richard Swedberg (eds.), *The Sociology of Economic Life*, 3rd edn., Westview Press (2011).

2. Clifford Geertz, "The bazaar economy: Information and search in peasant marketing," *The American Economic Review*, 68 (1978), 28.

3. M. Csíkszentmihályi, *Creativity: Flow and the Psychology of Discovery and Invention*, Harper Perennial, (1996), p. 39.

4. Plausible theory and a large following guarantee to make fame/fortune in academe. According to Paul Samuelson, in his presidential address to the American Economic Association in 1961: "In the long run the economic scholar works for the only coin worth having—our own applause."

5. For Natural Selection, Darwin figuratively portrayed the driving force: "The face of Nature may be compared to a yielding surface, with ten thousand sharp wedges packed close together and driven inwards by incessant blows, sometimes one wedge being struck, and then another with greater force" *On the Origin of Species* (1859).

6. William Gibson is reported as having said this during an interview on "Fresh Air," National Public Radio, August 31, 1993.

7. Douglass C. North, *Institutions, Institutional Change and Economic Performance*, Cambridge UP (1990), wherein he says: "Institutions are formed to reduce uncertainty in human exchange."

8. A. Gawande, "The Checklist: If something so simple can transform intensive care, what else can it do?" *The New Yorker*, December 10, 2007. See also *The Checklist Manifesto*, Picador (2011).

9. W. P. Bridges and R. L. Nelson, "Economic and sociological approaches to gender inequality in pay," in *The Sociology of Economic Life*, chapter 10.

10. For example, in the fierce competition between two airlines, consumers benefit from below-cost flights.

11. "Google claims $80 billion of economic impact on US economy," *Search Engine Journal*, July 3, 2012.

12. You may be twice as efficient as your colleagues, but can be still content with a 20% income difference, due to the partial appropriation effect.

13. J. Wallis, and D. North, *Measuring the Transaction Sector in the American Economy, 1870–1970*, Chicago University Press (1986).

14. Mark Granovetter, *The Impact of Social Structure on Economic Outcomes*, J. Econ. Perspectives 19, 33, 2005.

15. Brian Uzzi, "Social structure and competition in interfirm networks: The paradox of embeddedness," in *The Sociology of Economic Life*, chapter 12.

16. Brian Arthur, *The Nature of Technology: What It Is and How It Evolves*, Free Press (2009). Arthur posits that technologies can combine, be subdivided, and find new use in the economy.

17. The wonderful aspect of inefficiencies or non-zero margins was first emphasized by Tibor Scitovsky in "The benefits of asymmetric markets," *Journal of Economic Perspectives*, 4 (1990), 135.

18. Alexander Pope, *An Essay on Man* (1733).

19. L. Pecchi and G. Piga, *Revisiting Keynes: Economic Possibilities for Our Grandchildren*, MIT Press (2008).

20. Milton Friedman, *Capitalism and Freedom:* 40th edn., Chicago University Press (2002), in which he advocates that doctors licensing and FDA should be abolished.

21. A. Gawande, "Testing, testing," *The New Yorker*, December 2009: "The government never took over agriculture, but the government didn't leave it alone, either."

22. Similar points were made by Richard Dawkins when he compared bats and radar in *The Blind Watchmaker: Why the Evidence of Evolution Reveals a Universe without Design*, Norton (1996).

Chapter 8: Man and the Economy

1. David Ricardo, "On wages," in *On The Principles of Political Economy and Taxation* (1817), chapter 5.

2. Veblen, Scitovsky, Kahneman, Tversky, et al. pioneered the combining of human psychology with economics.

3. One of the leading members of the Chicago School, Gary Becker, is more explicit in promoting the notion that self-interest accounts for many human activities, even beyond the economy. *A Treatise on the Family*, Harvard University Press (1993).

4. Mainstream economics explains why people compete—fighting for self-interest—but has little to say about why people cooperate. See, however, new research by E. Fehr and S. Gaechter, "Cooperation and punishment in public goods experiments," *The American Economic Review*, 90 (2000), 980; E. Fehr and U. Fischbacher, "The nature of human altruism," *Nature*, 425, 785; H. Gintis and E. Fehr, "The social structure of cooperation and punishment," *Behavioral and Brain Sciences*, 35 (2012), 28.

5. See the pioneering work by Nobel laureate E. Ostrom, *Governing the Commons: The Evolution of Institutions for Collective Action*, Cambridge University Press (1990); see also the concept of "positive deviants" in Atul Gawande, *Better: A Surgeon's Notes on Performance*, Metropolitan Books (2007).

6. E. Ostrom *Governing the Commons*.

7. M. Ingram, "The Carr–Benkler wager and the peer-powered economy," *Gigaom*, May 9, 2012.

8. B. Perens, "The emerging economic paradigm of open source," *First Monday*, October 2005.

9. P. Gupta, "The pleasure principle: Not all products need to be painkillers," *Techcrunch*, January 9, 2011.

10. A quote attributed to Samuel Butler by D. N. Perkins: "The possibility of invention," in Robert J. Sternberg (ed.), *The Nature of Creativity*, Cambridge University Press (1988).

11. Clay Shirky, "Fame vs. fortune: Micropayments and free content." <http://www.shirky.com>.

12. Gandhi said it well: "Earth provides enough to satisfy every man's need, but not every man's greed."

13. Whale-watching tickets can be counted, but why people take the trouble and spend money to watch is more difficult to articulate.

14. See *Businessweek*, November 10, 1999, p. 52. For a more recent example, Steve Jobs did not ask consumers if they would like to have an iPhone, as the answer would be a resounding "No." See Adrienne Lafrance, "17 of the best Internet reactions to the original iPhone," *The Atlantic*, November 10, 2015.

15. B. Shneiderman, "Designers make data much easier to digest," *New York Times*, April 3, 2011.

16. Abraham H. Maslow, *Motivation and Personality*, Harper & Row (1970).

17. Maslow's hierarchy of needs is usually represented as a pyramid, with basic needs such as food, shelter at the bottom, and higher needs at the top. The pyramid shape is misleading, as the basic needs are smaller in number; but while there is no denying that they are more important, Maslow might have used it to emphasize what is important (basics). From the point of view of diversity, if the numbers are to be represented, we propose an inverted pyramid.

18. Maslow, *Motivation*, p. 7.

19. Michael Polanyi, *The Tacit Dimension*, University of Chicago Press (1966): "I shall reconsider human knowledge by starting from the fact that we can know more than we can tell." Polanyi's observations concerned science and technology, but now we extend a tacit dimension to the economy and society.

20. In Section 1.4 we mentioned that wants are horizontal and skills are vertical. While aggregate offers must be commensurate with consumers' wants in number, a single business must focus on its own niche. In the vertical direction, a firm can increase production, improve quality, and be more efficient, whereas a consumer in the horizontal direction will activate more and more wants that are either acquired or recommended.

21. S. Bhattacharya, "Personality changes throughout life," *New Scientist*, May 11, 2003.

22. For example, Adam Smith says: "Honour makes a great part of the reward of all honourable professions." *The Wealth of Nations*, vol. 1, chapter 10.

23. See, for example, John R. Anderson, *Cognitive Psychology and its Implications*, Worth Publishers, 7th edn. (2009).

24. Polanyi, *The Tacit Dimension*.

25. Max Boisot, *Information Space*, Routledge (2016); Max Boisot, Ian MacMillan, Kyeong S. Han, *Explorations in Information Space: Knowledge, Agents, and Organization*, Oxford University Press (2007).

26. Thorstein Veblen, *The Instinct of Workmanship and the State of Industrial Arts*, Forgotten Books (2015).

27. "IBM examines how inventors invent." <http://wired.com>, September 2003.

28. Abraham H. Maslow and Robert Frager, *Motivation and Personality*, 3rd edn., Longman (1987), p. 161.

29. See the debate on the importance of selection (Darwinian) compared to planning (Lamarckian). G. M. Hodgson and T. Knudsen, "In search of general evolutionary principles: Why Darwinism is too important to be left to the biologists," *Journal of Bioeconomics*, 10 (2008), 51–69. D. K. Simonton, *Origins of Genius: Darwinian Perspectives on Creativity*, Oxford University Press (1999). Robert Merton has also suggested that selection plays an important role with "unintended consequences of intended actions." R. K. Merton and E. Barber, *The Travels and Adventures of Serendipity: A Study in Sociological Semantics and the Sociology of Science*, Princeton University Press (2005).

30. "I think there is a world market for about five computers." Quote attributed to IBM's founder Thomas J. Watson (1943).

31. Mark Granovetter, "The impact of social structure on economic outcomes," *Journal of Economic Perspectives*, 19 (2005), 33–50.

32. Mark Granovetter, "The strength of weak ties," *American Journal of Sociology*, 78 (1973) 1360–80.

33. Becker, *A Treatise on the Family*. See also his Nobel Prize lecture (1992) on the economic way of looking at life.

34. This is related to Elinor Ostrom's concept of "thick rationality."

35. Marketing Guru Philip Kotler has said: "Good companies will meet needs; great companies will create markets." Ccited in *The 75 Greatest Management Decisions Ever Made*, MJF Books (2002), p. 37.

36. Michael Pollan, *The Omnivore's Dilemma: A Natural History of Four Meals*, Penguin (2007).

37. Susan Spano, "The Great Wall: Across China's great divide," *Los Angeles Times*, August 1, 2008. See also William Lindesay, *The Great Wall Revisited*, Harvard University Press (2008).

38. J. Brocke, "Volunteers in Japan give Mount Fuji a makeover," *New York Times*, December 7, 2003.

39. See <http://sustainablog.org/2009/04/pesticide-lobby-bugged-michelle-obamas-white-house-organic-garden>.

40. See <http://www.talkativeman.com/sam-walton-frugality-wal-mart/.>

41. Stephen A. Marglin, *The Dismal Science: How Thinking Like an Economist Undermines Community*, Harvard University Press (2010).

42. Bruno S. Frey, *Inspiring Economics: Human Motivation in Political Economy*, Edward Elgar (2001).

43. Richard Holloway, *Between the Monster and the Saint: Reflections on the Human Condition*, Canongate Books (2008); Daniel Kahneman, in his Nobel Prize lecture (2002), stated: "Some people were better than others, but the best were far from perfect and no one was simply bad.",.

44. The sixth US president John Quincy Adams (son of the second US president John Adams) is quoted as saying: "All men profess honesty as long as they can. To believe all men honest would be folly. To believe none so is something worse."

Chapter 9: Evolving Economy

1. Chris Anderson, "Tech is too cheap to meter: It's time to manage for abundance, not scarcity," *Wired Magazine*, June 2009.
2. J. Rosen, "The Web means the end of forgetting," *New York Times*, July 21, 2010.
3. Kevin Kelly, *New Rules for the New Economy: 10 Radical Strategies for a Connected World*, Penguin (1999).
4. "Surgery mix-ups surprisingly common." <http://edition.cnn.com/2010/HEALTH/10/18/health.surgery.mixups.common>.
5. Michael Lewis, *Next: The Future Just Happened*, Norton (2002).
6. Report in the Swiss newspaper *Matin Bleu*, July 19, 2007.
7. Katie Dean, "New spin on the music business," *Wired Magazine*, May 2004.
8. For example, the petition of candlemakers (historical) and many provisions of copyrights (current) impede innovation.
9. Catherine Arnst, "Health 2.0: Patients as partners," *Businessweek*, December 4, 2008.
10. Chinese doctors peddling medicine.
11. Atul Gawande, "America's epidemic of unnecessary care," *The New Yorker*, May 11, 2015.
12. See "Speeding the buy cycle," *Businessweek*, September 18, 2000.
13. Chris Anderson, *Free: How Today's Smartest Businesses Profit by Giving Something for Nothing*, Hachette Books (2010).
14. Frank Hahn, *On the Nature of Equilibrium in Economics*, Cambridge University Press (1973), pp. 14–15.
15. Traci Watson, "86 percent of Earth's species still unknown?" *National Geographic News*, August 25, 2011.
16. Jeremy B. Williams and Judith McNeill, "The current crisis in neoclassical economics and the case for an economic analysis based on sustainable development," *U21 Global Working Paper*, 001/2005.
17. M. Moss, "The hard sell on salt," *New York Times*, May 29, 2010.
18. This violates the Pareto principle of mainstream economics.
19. See, for example, "China cigarette order goes up in smoke," in which it is reported that officials were ordered to smoke at least 23,000 packs of cigarettes a year to help the local economy. <http://www.reuters.com/article/idUSPEK124354>.
20. Brian Arthur, "How fast is technology evolving?" *Scientific American*, February 1997.

21. Darwin proposed sexual selection at the same time as when he proposed Natural Selection.

22. Dawkins, *The Blind Watchmaker*.

23. Kelly, *New Rules for the New Economy*, chapter 6.

24. Stephen J. Gould *The Structure of Evolution Theory*, Belknap Press (2002).

Chapter 10: Paradigm Shift

1. Scitovsky, *Joyless Economy*, p. 176, where he emphasizes that it is diligent consumers' scrutiny that retains the price–quality correlation.

2. Krugman, "There'll always be a Soros."

3. William Heuslein, "The man who predicts the medals," *Forbes*, January 19, 2010.

4. Michael Lewis, *Moneyball: The Art of Winning an Unfair Game*, Norton (2004).

5. Atul Gawande, "University of Chicago medical school commencement address," *The New Yorker*, June 12, 2009.

6. Richard Van Noorden, "Physicists make 'weather forecasts' for economies," *Nature News*, February 23, 2015.

7. B. Lomborg, *The Skeptical Environmentalist: Measuring the Real State of the World*, Cambridge University Press (2001).

8. Lionel Robbins, *An Essay on the Nature and Significance of Economic Science*, Macmillan (1932).

9. Brad DeLong, "By the bags-of-flour standard, we are some 430 times wealthier than our typical rural ancestors of half a millennium ago," in Arnold Kling, *Learning Economics*, Xlibris Corporation (2004), p. 72.

10. Y.-C. Zhang, "Broader scopes of the reflexivity principle in the economy," *Journal of Economic Methodology*, 20 (2013), 446.

11. Kelly, "Opportunities before efficiencies," in *New Rules for the New Economy*, chapter 10.

12. K. A. Strassel, "Get your priorities right," *Wall Street Journal*, July 8, 2006.

13. A. M. Spence, "Job market signaling," *Quarterly Journal of Economics*, 87 (1973), 355–74.

14. Christian Arnsperger and Yanis Varoufakis, "What Is neoclassical economics," *Post-Autistic Economics Review*, 38 (2006).

15. Cliffe Leslie, "The known and unknown in the economic world," *Fortnightly Review*, June 1879, p. 934.

16. W. Brian Arthur, "The end of certainty in economics," in Diederik Aerts, Jan Broekaert, and Ernest Mathijs (eds.), *Einstein Meets Magritte: An Interdisciplinary Reflection on Science, Nature, Art, Human Action and Society*, Springer (1999), pp. 255–65.

17. Bertin Martens, *The Cognitive Mechanics of Economic Development and Institutional Change*, Routledge (2007).

18. J. M. Buchanan and V. J Vanberg, "The market as a creative process," *Economics and Philosophy*, 7 (1991), 167–86.
19. Saras D. Sarasvathy, *Effectuation: Elements of Entrepreneurial Expertise*, Edward Elgar (2009).
20. Y.-C. Zhang, "Broader scopes."
21. Constraint moving can be classified by the following categories: 1, deliberately want; 2, unplanned but pleasantly surprised; 3, saw it coming but resisted; 4 caught by surprise and did not have time to fight it off.

Index